DARWIN'S ENIGMA

FRED HARDING

"UK ENGLISH FORMAT"

COPYRIGHT

First published in UNITED KINGDOM

Copyright © Fred Harding

PICTURE CREDITS AND DISCLAIMER

Unless stated otherwise, the photos and artwork in this book were created by the author, including some he himself has modelled. Such pictures belong to Fred Harding and are copyright.

Third Party Photos and Drawings are acknowledged, permissions obtained and credits referenced. Some pictures originate from public domain or out of copyright sources.

Quotations referred to in this book are permitted under copyright law and "Fair Use" where a "reviewer may fairly cite largely from the original work, if his design be really and truly to use the passages for the purposes of fair and reasonable criticism."

Reviews, commentary and criticism are highly protected under copyright law. The "fair use" exemption to (U.S.) copyright law was created to allow things such as commentary, parody, news reporting, research and education about copyrighted works without the permission of the author.

With regard to pictures, "If you're using the images for the purposes of criticism or commentary, then it's usually Fair Use and you can use the images. For example, posting a picture of a cereal box alongside a review of that cereal would qualify as Fair Use of the image." Screenshot thumbnails (smaller regions of the main), also qualify as Fair Use.

CONTENTS

PREFACE

"There are one hundred and ninety-three living species of monkeys and apes. One hundred and ninety-two of them are covered with hair. The exception is a naked ape self-named Homo sapiens."
(Desmond Morris - The Naked Ape)

Approximately five thousand mammal species exist on earth and nearly all of them have a fur covering that covers almost their entire body. Desmond Morris, in his introduction to **The Naked Ape: A Zoologist's Study of the Human Animal**, remarked that of the 193 species of monkeys and apes, all but one is covered with hair. He identified Homo sapiens, the self-named ape, as the exception.

Morris was correct in his analogy. Among primates, humans are unique in having nearly naked skin. Every other member of our alleged extended family has a dense covering of fur - from the short, black pelage of the howler monkey to the flowing copper coat of the orangutan - as do most other mammals. The picture of my body on the cover of this book when compared to the gorilla says it all. My body is virtually devoid of hair. Yes, we humans do have hair on our heads, face (male) and pubic regions, but compared with our relatives, even the hairiest person is basically bare.

Until the late nineteenth century the reason for our naked almost hairless skin posed no problems and was not a cause for concern. This was because most people were Christians and believed in the Bible. They could recall the words of Jesus when he said, "Haven't you read," he replied, "*that at the beginning the Creator 'made them male and female...,*'" (Matthew 19:4) referring to the Genesis where it is written, "*So God created mankind in his own image, in the image of God he created them; male and female he created them.*" (Genesis 1:27) "*And the man and his wife were both naked and were not ashamed.*" (Genesis 2:25)

Then Victorian society was rocked by the announcement from part of the scientific community that man was not created in the image of God but was in fact, the image of an ape. The announcement came in 1863 when Thomas Henry Huxley (1825-1895) published a book entitled, **Evidence as to Man's Place in Nature** and in it he declared, "*Without question... [man's] early stages of development... [are] far nearer the apes, than apes are to the dog.*" [1]

Huxley having compared the adult anatomy of apes with man, asked the question, "*Is man so different from any of these apes that he must form an order by himself?*" [2]

In those few short words, Huxley had suggested that humans was not a unique creature made in the image of God but rather may in fact have kinship with the apes. "*It is quite certain that the ape which most nearly approaches man,*" he said, "*is either the Chimpanzee, or the Gorilla...*" [3]

Thomas Henry Huxley

Huxley's book summarized the many anatomical traits shared by humans and apes and asserted that such evidence supported the hypothesis that humans and apes had evolved from a recent common ancestor. Backed by this evidence, the book proposed to a wide readership that evolution applied as fully to man as to all other life. From this Huxley came to an earth shattering conclusion.

> *"But if man be separated by no greater structural barrier from the brutes than they are from each other-then it seems to follow that... there would be no rational ground for doubting that man might have originated... by the gradual modification of a man-like ape".* [4]

Huxley used the term "man-like ape" no less that 32 times in his book and he included the gorillas, gibbons, orangutans and chimpanzees under this label. His attempt to place man taxonomically with the higher apes shocked many Victorians, but then he appealed for support for his theory of the evolution of man and apes from a common ancestor, by saying:

> *"At the present moment there is but one hypothesis which has any scientific existence - that propounded by Mr. Darwin."* [5]

Of course the hypothesis that Huxley was referring to was that described in Darwin's book, **On the Origin of Species by Means of Natural Selection, or the Preservation of Favoured Races in the Struggle for Life.** published four years earlier in 1859 After the sixth edition the title was changed to simply **The Origin of Species**.

Darwin's book introduced the scientific theory that populations evolved over the course of generations through a process he called natural selection. However, if the reader was to look within its pages Darwin does not make any mention of the evolution of apes and humans from a common ancestor as Huxley had indicated. This was Huxley's conjecture. By appealing to a reputable scientist it gave the impression to the reader that his theory was one that Darwin himself supported. It is the same tactic that the leading proponent of evolution today, the atheist Richard Dawkins, uses to add credibility to his writings.

ALFRED WALLACE SHAKES SOME FEATHERS

Darwin was an eminent scientist, so anything he said was taken seriously and the evidence he presented for the change of species through natural selection generated considerable scientific, philosophical, and religious debate. However, the truth of the matter is that Darwin wrote his famous book under duress, because another British naturist had also discovered the mechanism of natural selection and was about to publish his findings before Darwin had published his. His name was Alfred Russel Wallace (1823-1913).

Wallace was considered the 19th century's leading expert on the geographical distribution of animal species and is sometimes called the "father of biogeography". It was while in Sarawak as the guest of James Brooke, a British adventurer whose exploits in the Malay Archipelago made him the first White Rajah of Sarawak, that he wrote his first paper that hinted at the theory of natural selection. The paper was published in 1855 and was entitled **On the law which has regulated the introduction of new species**. This was later known as the "Sarawak Law" and in it Wallace declared that, "*Every species has come into existence coincident both in space and time with closely allied species, connects together and renders intelligible a vast number of independent and hitherto unexplained facts.*" [6]

Alfred Wallace

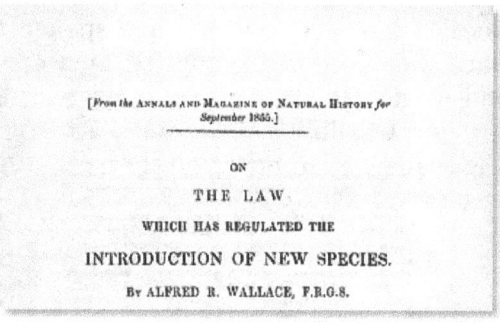

On the law which has regulated
the introduction of new species (1855)

Those words may sound nonsensical to most readers but what Wallace was suggesting was there appeared to be a close relationship and changes between species and an understanding of this could render a vast number of independent and hitherto unexplained facts intelligible. However, at this early stage he had no model of how those changes were actually achieved, but it is clear from the paper that he was thinking along the lines of Darwin's Galapogos islands discoveries that was published in **Origins of Species.**

> "*A country having species, genera, and whole families peculiar to it, will be the necessary result of its having been isolated for a long period, sufficient for many series of species to have been created on the type of pre-existing ones, which, as well as many of the earlier-formed species, have become extinct, and thus*

made the groups appear isolated. If in any case the antitype had an extensive range, two or more groups of species might have been formed, each varying from it in a different manner, and thus producing several representative or analogous groups." [7]

Wallace's "Sarawak Law" paper made such an impression on the famous geologist Charles Lyell (1797-1875), the foremost geologist of his day, that in November 1855, soon after reading it, he started a "species notebook" in which he began to seriously contemplate the implications of evolutionary change for the first time.

DARWIN'S DECEPTION

In April 1856 Lyell paid a visit to Charles Darwin at Down House. When Darwin explained his theory of natural selection to Lyell for the first time: a theory which Darwin had been working on, more or less in secret, for about 20 years, Lyell found himself in a dilemma. He was aware that Wallace's "Sarawak Law" too had spoken about a similar theory and that Wallace was close to finding the answer the mechanism involved. What should he do? In the end soon afterwards Lyell sent a letter to Darwin urging him to publish the theory *"lest someone beat him to it"*, no doubt having Wallace in mind.

Charles Darwin   Joseph Hooker   Charles Lyell

Without knowing why, in May 1856 Darwin heeded the advice of his friend and began to write a "sketch" of his ideas for publication. This "sketch" was abandoned in about October 1856 and Darwin instead began to write an extensive book about his theory of natural selection.

In February 1858 Wallace was suffering from malaria in the village of Dodinga on the remote Indonesian island of Halmahera when suddenly the idea of natural selection as the mechanism of evolutionary change occurred to him. As soon as he had sufficient strength he wrote a detailed essay explaining his theory and sent it together with a covering letter to Darwin, who he knew from previous correspondence was interested in the subject of evolution. He asked Darwin to pass the essay on to Charles Lyell if Darwin thought it was sufficiently interesting.

Wallace had as yet not corresponded with Lyell but in a previous correspondence Darwin mentioned that Lyell had found Wallace's 1855 paper noteworthy. Some people have suggested that the reason why Wallace wanted Lyell to read his essay was because it was written as an argument against the anti-evolutionary views in Lyell's book, **Principles of Geology**.

Darwin was horrified when he received Wallace's letter because he had discovered natural selection many years earlier but had not published his findings in any scientific paper. Wallace it seemed would beat him to the glory of the discovery and steal his thunder. Darwin immediately appealed to his influential friends Lyell and Joseph Hooker (1817-1911), the latter being the Director of the Royal Botanical Gardens, Kew, and one of the greatest British botanists and explorers of the 19th century, for advice on what to do. Lyell and Hooker thought the best solution was to present Wallace's essay along with two unpublished excerpts from Darwin's writings on the subject, to a meeting of the Linnean Society of London on 1 July 1858. Darwin agreed.

These documents were published together in the Society's journal on 20 August of the same year as the paper **On the Tendency of Species to Form Varieties; And On the Perpetuation of Varieties and Species by Natural Means of Selection**. Darwin's contributions were placed before Wallace's essay, thus emphasising Darwin's priority to the idea. This was the first part of the deception. The next was far worse. The paper was published without Wallace's knowledge or consent. Wallace later remarked that the paper "*was printed without my knowledge, and of course without any correction of proofs*", contradicting Lyell and Hooker's statement in their introduction to the joint papers that "*both authors...[have]...unreservedly placed their papers in our hands*". [8]

Charles Darwin's son, Francis, acknowledged in **More Letters of Charles Darwin** that had Wallace published his paper directly, it would have been Wallace and not his father who would have been credited with the discovery of natural selection.

"Wallace was the one who had the paper ready for publication, and if he'd sent it directly to a journal it would have been published and natural selection would have been Wallace's discovery." [9]

Dr George Beccaloni, a curator at the Natural History Museum, said that this was a crime because they did this without Wallace's knowledge.

"Dr Beccaloni said there was no better time to recognise that, while Darwin had been working on the theory of natural selection for many years, Wallace had also been toiling away on the same idea. The two men made breakthroughs independently but Wallace was the first to write an explanation down in an essay. According to Dr Beccaloni, however, Wallace's mistake was sending the essay from Indonesia -where he had been researching for eight years - to Darwin, in the course of a correspondence between the two men. The curator says Wallace had no idea his scientific pen pal been working on the same theory." [10]

Neither author was present at the presentation at the Linnean Society. Darwin was attending the funeral of his son, and Wallace was still in Borneo. The meeting was chaired by the President of the society, Thomas Bell, who had written up the description of Darwin's reptile specimens from the Beagle expedition. About thirty members of the society were present at the meeting and it just so happened that Wallace's natural history agent Samuel Stevens was also attending. It was through him that Wallace learned about the publication of his paper.

Shortly after the Linnean Society meeting Darwin wrote to Wallace to explain what had occurred but by then Wallace already knew. Wallace, ever humble, accepted the arrangement after the fact, happy that he had been included at all, and never expressed public or private bitterness about the affair. Although Lyell and Hooker's arrangement relegated Wallace to the position of co-discoverer, it was Darwin who gained the most recognition for the theory that both had discovered.

"Dr Beccaloni said the actions of Lyell and Hooker were "pretty morally reprehensible... If Wallace had not sent it to Darwin and [instead] sent it to a scientific journal, then he would have had priority and we would be talking about Wallaceism, not Darwinism"." [11]

In the following year after the publication their joint paper, Darwin made sure that his recognition as being the primary discoverer of natural selection was certain with the publication of his famous **Origins** book.

Darwin had for many years been working on a massive book on natural selection, which he called his "big book". But Wallace's work had now forced his hand. For thirteen months Darwin struggled to produce an abstract of the "big book", suffering from ill health in the process but he received constant encouragement from his scientific friends. Finally, Lyell arranged to have the abstract published by John Murray in November 1859. It was called *On the Origin of Species by Means of Natural Selection, or the Preservation of Favoured Races in the Struggle for Life*. It was an outstanding success.

As the acceptance of natural selection gathered pace and more and more naturalists joined Darwin's band of rebels against the Creationist status quo, Wallace remained in the background lending his support. Hence, in many accounts of the development of evolutionary theory, Wallace is mentioned only in passing as being the stimulus to the publication of Darwin's own theory.

However, Wallace was developing his own distinct evolutionary views on natural selection which diverged from Darwin's. His was a purer untainted version of natural selection while Darwin's theory had embedded within it the long discredited transmutation of the species that his grandfather Erasmus had once advocated. Their differences became apparent when the question of Man's hairlessness came to be discussed. It was Wallace's view that it was absolutely certain that natural selection could not have been responsible for humans becoming hairless.

> **"Man's Naked Skin could not have been produced by Natural Selection.** *It seems to me, then, to be absolutely certain, that' Natural Selection" could not have produced man's hairless body by the accumulation of variations from a hairy ancestor. The evidence all goes to show that such variations could not have been useful, but must, on the contrary, have been to some extent hurtful."* [12] [bold mine]

The issue of our naked skin and almost human hairlessness was to become Darwin's enigma, a problem that he never was able to resolve. Nor have any naturalist scientist since, although it has not been for the want of trying. It is the objective of this book to take a honest look at all the theories proposed by Darwin and those who followed him to see if there is indeed a solution through a naturalist mechanism that could account for the anomolity of Man's almost hairless condition. But if there is no solution that can be found, then we will have to face one unavoidable conclusion. Man did not evolve from a common ape-like ancestor as is taught today but he was created through the activities of Intelligent Design in an act of Creation.

CHAPTER 1
NATURAL SELECTION IN CONFLICT

"The wonders wrought by artificial selection in the breeding of domestic animals and cultivated plants are such that one might well have attributed great results to the exercise of a similar selection by Nature through countless ages, could any such process be detected."
(The Triumph of Darwinism - 1877)

Charles Darwin   Alfred Wallace

TWO VERSIONS OF NATURAL SELECTION | THE TRANSMUTATION OF SPECIES | LAMARK GRABS THE LIMELIGHT | DARWIN BELITTLES LAMARCK | VESTIGES PAVES THE WAY FOR EVOLUTION | DARWIN AND VESTIGES | THE RISE OF DARWIN'S EVOLUTION | WALLACE AND LYELL EMBRACED INTELLIGENT DESIGN

In Victorian times Christendom taught an erroneous doctrine that was not found in the Bible. It was called the "Immutability (or fixity) of the Species" and it was the idea that each individual species on the planet was specially created by God and could never fundamentally change. It was Thomas Aquinas (1225-1274) an Italian Dominican friar and priest who more than any other who introduced this idea into the Church. He is honoured as a saint by the Catholic Church and is held to be the model teacher for those studying for the priesthood, and indeed the highest expression of both natural reason and speculative theology.

In modern times, under papal directives, the study of Aquinas works was long used as a core of the required program of study for those seeking ordination as priests or deacons, as well as for those in religious formation and for other students of the sacred disciplines (Catholic philosophy, theology, history, liturgy, and canon law). [1] Aquinas is honoured as a Doctor of the Church and is considered the Church's greatest theologian and philosopher.

There is no doubt that the philosophical thoughts of Aquinas has exerted enormous influence on Christian theology, especially that of the Catholic Church and has extended to Western philosophy in general too. He stands as a vehicle and modifier of Aristotelianism and Neoplatonism and that was the problem. He and others like him blended Aristotelian philosophy with Catholic Christianity so that over time, ideas that were not scriptural found their way into the doctrines of the Church and were given a biblical facade of authority. One of these was the doctrine of the immutability of species aforementioned.

By the seventeenth century, this doctrine had become so linked to the belief in separate creation as taught in Genesis that the Swedish botanist Carolus Linnaeus (1707 - 1778), famous for laying the foundations of our modern biological classification system, could say with sincere conviction: "We count as many species as there were created forms in the beginning." [2] So it was that at the dawn of the nineteenth century the doctrine of immutability of species was considered Holy Writ and most naturalists accepted this erroneous idea in a similar vein. [3] Hence, when Darwin wrote his famous book *On the Origin of Species* he said in it that all the most eminent paleontologists and all the greatest geologists have unanimously, often vehemently, maintained the immutability of species. [4]

Darwin did not wish to contradict such great authorities of whom he admired but both he and Wallace could see that man was able to make changes to existing species of both plants and animals through a process called selective breeding or "Artificial Selection" as Darwin called it. Both naturalists wondered that if man could change an existing species into a different form, then could not nature do the same they asked. Through their research and observations both Darwin and Wallace could show conclusively that nature could indeed make similar changes and thus, the theory of Natural Selection was born.

TWO VERSIONS OF NATURAL SELECTION

Darwin and Wallace to their credit corrected a popular misunderstanding and the Church was proved wrong but this did not mean what was written in the Bible was incorrect. It was only those who had formulated doctrines outside the Biblical narrative that were in error. But just as one misconception had been proved wrong, in the Darwinian camp a new misconception was beginning to emerge that would take it's place. Originally called the **Transmutation of Species**, it is better known today as evolution.

Let us make this abundantly clear, contrary to popular belief natural selection is not evolution. It is really nothing more than selective breeding where nature and not the intervention of a human breeder is involved. Both Darwin and Wallace said as much and this can be readily proved because despite all the intellectual and technological skills that man has at his disposal, when it comes to selective breeding he has never been able to transform an animal like a dog into anything else other than a dog. Although a dog may have changed shape, size, colour or even temperament through artificial selection a dog will always remain a dog regardless of how much selective breeding has taken place. Are we therefore to believe that what scientists cannot do today with all their knowledge and skills that somehow nature could achieve through random chance? That would stretch credibility to it's limits if this was so.

Wallace unlike Darwin could see that the accumulation of variations through natural selection from a hairy ancestor could not account for man's own hairless naked skin. According to Wallace all the evidence showed that such variations could not have been useful, but must have been to some extent damaging. This is exactly what George Cuvier (1769-1832) had concluded only decades before. Cuvier was the major figure in natural sciences research in the early nineteenth century. It was he who was instrumental in establishing the fields of comparative anatomy and palaeontology through his work in comparing living animals with fossils. Cuvier was therefore well qualified to say that for any change in an organism's anatomy it would have rendered it unable to survive. Cuvier saw organisms as integrated wholes, in which each part's form and function were integrated into the entire body. No part, he said, could be modified without impairing this functional integration.

It is because of his understanding of animal anatomy and physiology, that Cuvier strongly objected to any notion of evolution that some scientists were proposing in his day, which involved the gradual transmutation of one form into another, a theory known in his day as the **Transmutation of Species**.

THE TRANSMUTATION OF SPECIES

When I speak of Natural Selection not being evolution, I do so on the basis of the theory that Wallace formulated and not that of Darwin. While it is true that Darwin brought the concept of natural selection to public attention, his version had been corrupted by the merging of another theory, namely the **Transmutation of Species**. I know that is hard to believe but as we shall see shortly, Darwin admits it. How did this happen? When Wallace discovered natural selection he was not hampered by any preconceived ideas. His theory was based purely on careful field observations of organisms and their environment. Darwin on the other hand was not so fortunate. His grandfather Erasmus Darwin (1731-1802) cast a long shadow over him and many, if not most, of his major ideas are found in his grandfather's book *Zoonomia or the Laws of Organic Life* that was published in 1794. As Russell Grig in his article *Darwinism: it was all in the family* says:

> "*Almost every topic discussed, and example given, in Zoonomia reappears in Charles's Origin. In fact, all but one of Charles's books have their counterpart in a chapter of Zoonomia or an essay-note to one of Erasmus's poems. And Charles's own copies of Zoonomia and The Botanic Garden are extensively marked and annotated.*" [5]

Erasmus's tome was a two-volume medical work that dealt with pathology, anatomy, psychology, and the functioning of the body. Erasmus thought that all life had evolved from one common ancestor which over time branched off into all the species we see today. He thought this transmutation of species was driven by competition and sexual selection, the very theories that Charles would later pen in his famous books. This can hardly be a coincidence. Erasmus was also strongly anti-Church, and included Credulity, Superstitious Hope, and the Fear of Hell in his catalogue of diseases. So when Charles later denounced his belief in God, it would not have caused him much anxiety. It was in the family.

LAMARK GRABS THE LIMELIGHT

Just before Erasmus Darwin died, a French naturalist named Jean-Baptiste Lamarck stole his thunder. It was he that established a theory on evolution based upon the transmutation of species along the lines that Erasmus postulated, but with some perceived evidence to support the theory.

Lamarck published a series of books on invertebrate zoology and palaeontology. Of these, *Philosophie zoologique*, published in 1809, most clearly states his theories of evolution although eight years earlier he had made his thoughts known in a paper. In that paper he put forward his **Theory of Inheritance of Acquired Characteristics**. That theory goes like this. If an organism changes during life in order to adapt to its environment, those changes are passed on to its offspring.

Lamarck had been struck by the similarities of many of the animals he studied, and was impressed too by the burgeoning fossil record. It led him to argue that life was not fixed. When environments changed he argued, organisms had to change their behaviour in order to survive. This sounds like the survival of the fittest that Darwin would later promote as his theory. If we compare what Lamarck said above to what Darwin believed, the similarities cannot go unnoticed.

Jean-Baptiste Lamarck   Erasmus Darwin

Darwin believed that nature selected only those individuals that had favourable variations and thereby they would have a competitive advantage over the others. This process is what Darwin called **natural selection**. It was said that the members of any particular species could develop and survive only when they were able to adapt themselves to the changed environmental conditions, by virtue of their favourable variations. Such variations were inherited and after a number of generations these variations became so prominent that a new species formed. Thus, new species develop from the existing ones in a slow and gradual way, a process Darwin called **Speciation**. Today we have a new name for this term. It is called **Macroevolution**.

Although the outcome is described as the same, the mechanisms between what Darwin proposed and what Larmarck wrote about was different. Darwin believed that while variations were the raw materials for evolution, natural selection was the force responsible. According to Lamarck everything could be explained when characteristics acquired by an organism is transmitted by heredity to the next generation. In every generation, fresh characteristics are acquired. With the result, after many generations, the changes accumulated to such an extent that the species becomes modified into a new one. Thus according to Lamarck this then was how one species could be transmuted into another. This was evolution, the transmutation of species.

Lamarck postulated that if an animal began to use an organ more than they had in the past, it would increase in its lifetime. Hence, if a giraffe stretched its neck for leaves, for example, a "nervous fluid" would flow into its neck and make it longer. Its offspring would inherit the longer neck, and continued stretching would make it longer still over several generations. Likewise, Lamarck believed that elephants all used to have short trunks. When there was no food or water that they could reach with their short trunks, they stretched their trunks to reach the water and branches, and their offspring inherited long trunks. Meanwhile organs that organisms stopped using would shrink and finally disappear.

Similarly, Lamarck interpreted the absence of limbs in snakes as an evolutionary change. He suggested that originally the ancestors of snakes had well-developed limbs but they faced the problem of increased predation. In order to protect themselves, snakes started crawling on the ground into crevices and holes. Continuous disuse of limbs resulted in limbs becoming shorter and shorter and finally generations later they had disappeared altogether.

August Weismann, a German biologist, tested Lamarck's theory by cutting off the tails of mice generation after generation. If Lamarck's theory was correct, the subsequent generations should have developed shorter and shorter tails. No such shortening of tails were observed. Even after a very large number of generations had passed mice continued to develop tails of the same length. Lamarck's mechanism of evolution, the transmutation of species, did not stand up to scientific scrutiny, but the basic idea of evolution as a concept captured the minds of many secular scholars. The seeds had been sown from which Darwin would later capitalise.

DARWIN BELITTLES LAMARCK

In public Charles Darwin paid lip service to Lamarck. "He wrote in 1861: "*Lamarck was the first man whose conclusions on the subject excited much attention. This justly celebrated naturalist first published his views in 1801 ... he first did the eminent service of arousing attention to the probability of all changes in the organic, as well as in the inorganic world, being the result of law, and not of miraculous interposition.*"

In private though Darwin was not so amenable and with good reason. He believed that Lamarck's ideas had been stolen from his grandfather. As the Museum of California Museum of Palaeontology, Berkeley website says:

> "*Several other scientists of the day, including Erasmus Darwin, subscribed to the theory of use and disuse - in fact, Erasmus Darwin's evolutionary theory is so close to Lamarck's in many respects that it is surprising that, as far as is known now, the two men were unaware of each other's work.*" [6]

Erasmus may have been unaware of Lamarck's work but Charles certainly was not. Furthermore, Charles was to use much of his grandfather's work in his own writings.

> "*It is interesting to note*", says the Museum of California Museum of Palaeontology. "*that Lamarck cited in support of his theory of evolution many of the same lines of evidence that Darwin was to use in the Origin of Species. Lamarck's Philosophie zoologique mentions the great variety of animal and plant forms produced under human cultivation (Lamarck even anticipated Darwin in mentioning fantail pigeons!); the presence of vestigial, non-functional structures in many animals; and the presence of embryonic structures that have no counterpart in the adult.*" [7]

No wonder Charles was angry with Lamarck and privately he did all he could to discredit the French naturalist. In a letter that he wrote in 1844 to his friend Joseph Hooker (1817 - 1911) in which he talked about about evolution he said, "*With respect to books on this subject, I do not know any systematical ones, except Lamarck's, which is veritable rubbish*". [8]

In another letter to Thomas Huxley in the same year, Darwin insinuated that Lamarck may have taken his ideas from his grandfather without giving him proper credit. [9]

Of all the references to Lamarck in his private letters, Darwin does not speak favourably of him or his work. It is clear, in view of what Darwin would publish and plagiarise later, that he wanted to disassociate himself from Lamarck's evolution philosophy. Hence words such as "absurd", "nonsense" and "rubbish" permeated his letters to his colleagues about Lamarck's transmutation of species theory.

However, when it came to his published books, Darwin refrained from making derogatory remarks. Instead, he avoids commenting on Lamarck altogether. In the he mentions Lamarck only once, a relatively trivial point about analogical resemblance. In **Descent of Man** too, Lamarck is only mentioned once and as for **Variation of Animals and Plants Under Domestication** he is not mentioned at all. In all of Darwin's publications, Lamarck is either ridiculed or ignored.

VESTIGES PAVES THE WAY FOR EVOLUTION

In 1844, the same year that Darwin was writing to Hooker describing Lamarck's book as being rubbish, another book appeared out of nowhere. Called **Vestiges of the Natural History of Creation** it was written by Robert Chambers (1802-1871) a Scottish publisher, geologist, evolutionary thinker, author and journal editor who, like his elder brother and business partner William Chambers, was highly influential in mid-19th century scientific and political circles. The book had been published anonymously because what it contained was so controversial that Chambers did not wish to acknowledge it as being his own and it was only after his death that it was learned it was he who had written it.

VESTIGES

OF

THE NATURAL HISTORY

OF

CREATION.

WITH A SEQUEL.

NEW YORK:
HARPER & BROTHERS, PUBLISHERS,
329 & 331 PEARL STREET,
FRANKLIN SQUARE.

During the 1830s, Chambers had taken a particularly keen interest in the then rapidly expanding field of geology, and he was elected a fellow of the Geological Society of London in 1844. Prior to this, he was elected a member of the Royal Society of Edinburgh in 1840, which connected him through correspondence to numerous scientific men.

The reason for Chambers' anonymity was clear enough as soon as one began reading the text. The book combined stellar evolution with progressive transmutation of species in the same spirit as the late Lamarck. By the time the book was written Lamarck had long been discredited among intellectuals, and evolutionary (or development) theories were exceedingly unpopular, except among political radicals, materialists, and atheists. Chambers, however, tried to explicitly distance his own theory from that of Lamarck's by denying Lamarck's evolutionary mechanism any plausibility.

Writing in his book Chambers said:

> "*Now it is possible that wants and the exercise of faculties have entered in some manner into the production of the phenomena which we have been considering; but certainly not in the way suggested by Lamarck, whose whole notion is obviously so inadequate to account for the rise of the organic kingdoms, that we only can place it with pity among the follies of the wise.*" [10]

Chambers was certainly aware of the storm that would probably be raised at the time by his treatment of the subject, and most importantly he did not wish to get his and his brother's publishing firm involved in any kind of scandal that could potentially ruin or severely impact their business venture. The arrangements for publication, therefore, were made through a friend named Alexander Ireland, of Manchester.

To further prevent the possibility of any unwanted revelations, Chambers only disclosed the secret to four people: his wife, his brother William, Alexander Ireland, and George Combe's nephew, Robert Cox. [11] All correspondence to and from Chambers passed through Ireland's hands first, and all letters and manuscripts were dutifully transcribed in Mrs. Chamber's hand to prevent the possibility of anyone recognising Robert's handwriting.

Vestiges of the Natural History of Creation caused a sensation when it was published and it quickly went through a number of new editions. It was initially well received by polite Victorian society and became an international bestseller, but its unorthodox themes contradicted the natural theology fashionable at the time and were reviled by clergymen -

and subsequently by scientists who readily found fault with its amateurish deficiencies. Even so it was so popular that Prince Albert read it aloud to Queen Victoria in 1845.

Vestiges brought widespread discussion of evolution out onto the streets and gutter presses and into the drawing rooms of respectable men and women. It's influence was such that it paved the way for acceptance of Darwin's own theory of evolution when his work was finally published in 1859. Darwin later remarked that Vestiges was important in preparing many people to accept his own theory of evolution.

John van Wyhe, Professorial Fellow of Charles Darwin University, writes concerning the book:

> "*Reading the book in a post-Darwinian world often leads to the skewed representation of Vestiges as a flawed precursor of Darwin's Origin of Species (1859). However, during the 1840s and 1850s Vestiges was the only 'evolution' book readers in the English speaking world were familiar with. Although much of the critical invective directed against the book focused on the issue of speciation - readers of Vestiges found a grand tale of the "development" or progress of nature from swirling clouds of interstellar gas, to the geological ages of the Earth, to the increasing complexity of organic forms and the improvement of man. The "development" narrative of Vestiges is one modern readers may find quite familiar - but it was just this that was so odious - so shocking- to many Victorian readers.*" [12]

DARWIN AND VESTIGES

Darwin read *Vestiges of the Natural History of Creation* in November 1844 finding that it drew on some of the lines of evidence he himself was putting together, and introduced questions that had to be dealt with. [13] His friend Hooker was delighted with Vestiges, because of the multiplicity of facts the author brought together, although he did not agree with the author's conclusions. [14] In April 1847, after meeting Chambers and then subsequently receiving a presentation of *Vestiges*, Darwin became convinced that Chambers had been the author. [15]

In his introduction to *On the Origin of Species*, published in 1859, Darwin assumed that his readers were aware of the book so he began his tome by identifying what he felt was one of its gravest deficiencies with regards to its theory of biological evolution:

> "*The author of the 'Vestiges of Creation' would, I presume, say that, after a certain unknown number of generations, some bird*

had given birth to a woodpecker, and some plant to the mistletoe, and that these had been produced perfect as we now see them; but this assumption seems to me to be no explanation, for it leaves the case of the coadaptations of organic beings to each other and to their physical conditions of life, untouched and unexplained." [16]

Chambers took the publication of the Darwin's book as an opportunity to release a new edition of *Vestiges* in 1860 and responded to Darwin's comments saying that he had been misunderstood. Chambers wrote:

"It seems to the author that Mr. Darwin has only been enabled by his infinitely superior knowledge to point out a principle in what may be called practical animal life, which appears capable of bringing about the modifications theoretically assumed in the earlier work. His book, in no essential respect, contradicts the present: on the contrary...it expresses substantially the same general ideas." [17]

Chambers concludes that *"The difference seems to be in words, not in facts or effects."*

It is more than likely that Darwin had read Chambers'comments, because he removed the offending passage from the 3rd edition of the Origin (1861) and all subsequent editions. In a historical sketch, newly added to the 3rd edition, Darwin softened his language a bit:

"The author apparently believes that organisation progresses by sudden leaps, but that the effects produced by the conditions of life are gradual. He argues with much force on general grounds that species are not immutable productions. But I cannot see how the two supposed "impulses" account in a scientific sense for the numerous and beautiful co-adaptations which we see throughout nature; I cannot see that we thus gain any insight how, for instance, a woodpecker has become adapted to its peculiar habits of life. The work, from its powerful and brilliant style, though displaying in the earlier editions little accurate knowledge and a great want of scientific caution, immediately had a very wide circulation." [18]

Darwin even suggested that Chambers' book helped pave the way for the publication of his theory of evolution by natural selection.

"In my opinion it has done excellent service in this country in calling attention to the subject, in removing prejudice, and in thus preparing the ground for the reception of analogous views."[19]

The idea of evolution was now in the public arena and Darwin was ready for the challenge. Capitilising on the popularity of *Vestiges*, he was going to introduce the transmutation of the species that his grandfather had promoted and which he believed that Lamarck had stolen.

THE RISE OF DARWIN'S EVOLUTION

Darwin had in his possession the works of his grandfather Erasmus, and he kept four notebooks that were entirely devoted to research on the transmutation of species. In the *First Notebook on Transmutation of Species* he uses his grandfather's book "Zoonomia" as a guide and added notes and corrections to what Erasmus had written. This book was commenced about July, 1837 and probably ended in February of the following year.

When Wallace wrote to Darwin telling him about his theory, for a moment Darwin panicked. Someone had beaten him to his theory. So he had to work fast to get his work out before Wallace produced his tome. In one of his letters published in 1887 by his son Francis (1848 - 925) Darwin wrote:

> "*In September 1858 I set to work by the strong advice of Lyell and Hooker to prepare a volume on the* **transmutation of species**, *but was often interrupted by ill-health, and short visits to Dr. Lane's delightful hydropathic establishment at Moor Park. I abstracted the MS. begun on a much larger scale in 1856, and completed the volume on the same reduced scale. It cost me thirteen months and ten days' hard labour. It was published under the title of the 'Origin of Species,' in November 1859. Though considerably added to and corrected in the later editions, it has remained substantially the same book.*" [20] [bold mine]

Here we can have it! Darwin admitted that his book was really about the transmutation of the species and while it also contained his famous theory of natural selection, the theory was in fact a merging of the two. That is why in a letter to his friend Asa Gray, (1810 - 1888), considered the most important American botanist of the 19th century, he criticised a reviewer W. Hopkins who lumped Darwinian evolution with that of Lamarck's transformism, by saying that, "*he does not in the least appreciate the difference in my views and Lamarck's*". [21]

The problem was that even one of his closest friends Charles Lyell could see the similarity too. Darwin's daughter Henrietta (Etty) wrote to her father: "*Is it fair that Lyell always calls your theory a modification of Lamarck's?*" [22]

In 1863 responding to his daughter's observation Darwin wrote to Lyell expressing his anger at Lyell's assertion that *Origin of Species* was only a modification of Lamarck's views. In the letter Darwin argued that it was true that Larmarck had preceded him in the idea that species evolve, but he pointed out that Plato, Buffon and his grandfather, and many others had similar views.

Darwin wrote to Hooker on 13th March, 1863 about Lyell's comments saying that, "*I have grumbled a bit in my answer to him at his always classing my work as a modification of Lamarck's, which it is no more than any author who did not believe in immutability of species, and did believe in descent.*" [23]

WALLACE AND LYELL EMBRACED INTELLIGENT DESIGN

Darwin had forgotten that Lyell in his *Principles of Geology* (1830-1833) had criticised and dismissed Lamarck's theory of evolution, ie the transmutation of species, so Lyell was well acquainted with Lamarck's theory. When Lyell saw the similarities with what Darwin had written in *Origins* and what Lamarck had written Lyell was understandably concerned. Although as a friend Lyell had supported Darwin, eventually that support waned. Lyell could not bring himself to accept what appeared to be the same Lamarckian theory found in Darwin's book that he had fought so hard to discredit. In the end, Lyell turned away from Darwin's theory of natural selection with the transmutation of the species embedded inside and endorsed instead the purer untainted natural selection that Wallace advocated.

Edward Larson in his book *Evolution: The Remarkable History of Scientific Theory* published in 2004 says it all.

> "*The drift from orthodoxy of such professed Darwinists as Wallace and Haeckel shows that, by the end of the nineteenth century, Darwinism was on the ropes. Although Wallace relied more heavily on the natural selection of inborn variations to account for the normal course of evolution than even Darwin (who supplemented it with notions of acquired characteristics and correlated growth), he could not conceive of that process producing the great leaps forward represented by the first appearance of matter, life, animals, and humans. These steps*

required the intervention of an "Overruling Intelligence," he believed. Particularly, he viewed the human mind as so far superior to those of any other animals in ways not useful in the struggle for existence (such as moral reasoning and mathematical genius) that it could not have evolved in a Darwinian fashion. "Some of the greatest upholders of the theory of natural selection admit that these higher facilities cannot have been developed through its agency," Wallace asserted, citing Weismann and Huxley as examples. He could have added Lyell and Gray, as well." [24]

One part of Wallace's remarkable life and career has been completely ignored: His embrace of intelligent design. Wallace's belief in intelligent design was launched in an essay published in the Quarterly Review in 1869 where he called upon an "Overruling Intelligence" to account for the mind of man, an idea Wallace continued to develop in **Darwinism** (1889) and **Man's Place in the Universe** (1903). **The World of Life** supplied a fitting culmination to Wallace's life's work.

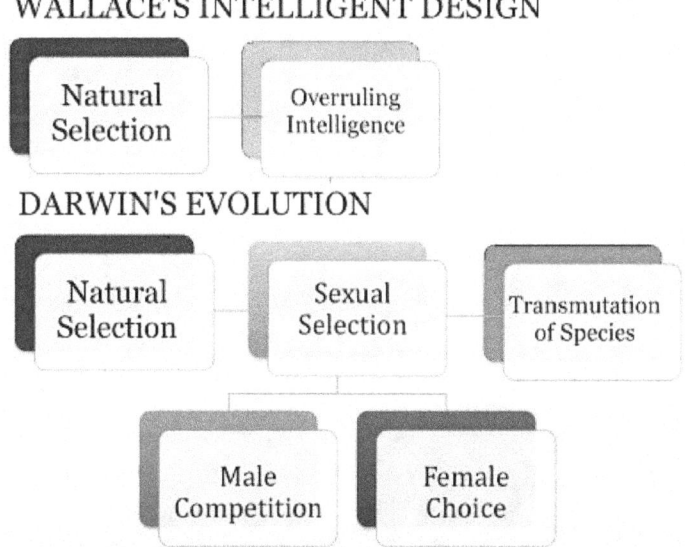

WALLACE'S INTELLIGENT DESIGN

DARWIN'S EVOLUTION

Professor Michael Flannery, author of the acclaimed biography, **Alfred Russel Wallace: A Rediscovered Life** said that both Wallace and Darwin were committed to science, but their conceptions were dramatically different. For Wallace science was simply the search for truth in the natural world but for Darwin science was one that invoked only natural processes without the hand of a Creator being involved. [25]

As for Lyell he wrote to Darwin making it clear that he agreed with Wallace about the overuling intelligence guiding the forces of the laws of nature.

> "*I rather hail Wallace's suggestion that there may be a Supreme Will and Power which may not abdicate its functions of interference, but may guide the forces and laws of Nature.*" [26]

If Darwin had any doubts about Lyell's support of his theory, he had none now. Lyell supported Wallace and Intelligent Design.

Wallace could say with conviction that man's nakedness could not have been the result of natural selection and he was right but Darwin had another trick up his sleeve. He had a second theory, which he called **Sexual Selection** and he thought that in this he had a solution for human hairlessness.

Chapter 2
DARWIN'S SECOND THEORY

"The evolution of near nakedness in the human species has been accounted for by a series of myths which owe more to the predilections of their creators than to the available evidence."
(Professor Francis Ebling,
"Journal of Human Evolution", January 1985)

THE CARD UP DARWIN'S SLEEVE | SEXUAL SELECTION IS PUBLISHED | THE TWO WAYS OF SEXUAL SELECTION | THE PEACOCK PROOF DISCREDITED | HAIRLESSNESS AND SEXUAL SELECTION | SEXUAL SELECTION IS REJECTED

While Darwin acknowledged that "*Man differs conspicuously from all the other Primates in being almost naked*", he was convinced that Man had descended from a hairy common ancestor. He had observed for example that "*a few short straggling hairs are found over the greater part of the body in the man, and fine down on that of the woman*". From this Darwin conjectured:

"There can be little doubt that the hairs thus scattered over the body are the rudiments of the uniform hairy coat of the lower animals." [1]

Wallace completely disagreed with Darwin over this matter and said that all the evidence showed that the accumulation of variations through natural selection from a hairy ancestor could not account for man's hairless naked skin. Wallace argued that natural selection operating repeatedly on random genetic changes over vast periods of time to fine-tune biochemical systems could not account for man's naked hairless condition. The incremental small variations resulting from the random genetic changes wrought by nature could not have been useful until the final result had been attained. He knew as did Darwin that most if not all the variations (genetic mutations) would have been harmful and would have been discarded by natural selection along the way. [2]

In *Origins* Darwin appears to agree with Wallace that natural selection could not have been the cause for the denudation of Man's body. Darwin talked about the Indian elephant as an example to demonstrate this.

"Elephants and rhinoceroses are almost hairless; and as certain extinct species, which formerly lived under an Arctic climate, were covered with long wool or hair, it would almost appear as if the existing species of both genera had lost their hairy covering from exposure to heat. This appears the more probable, as the elephants in India which live on elevated and cool districts are more hairy (87. Owen, 'Anatomy of Vertebrates,' vol. iii. p. 619.) than those on the lowlands. May we then infer that man became divested of hair from having aboriginally inhabited some tropical land?" [3]

At first glance from what Darwin has said that one might easily conclude that Man had become hairless because of living in the tropics of Africa and therefore through the processes of natural selection, man had lost his hair because the environment in which he lived was so beneficial that hair was not needed. Darwin even suggested because the hair was chiefly retained in the male sex on the chest and face, and in both sexes at the junction of all four limbs with the trunk that this would favour this conclusion, on the assumption that the hair was lost before man became erect; for the parts which now retain most hair would then have been most protected from the heat of the sun. However, in reality Darwin could see, like Wallace, that hair loss through natural selection was not the answer.

"The loss of hair is an inconvenience and probably an injury to man, even in a hot climate, for he is thus exposed to the scorching of the sun, and to sudden chills, especially during wet weather. As Mr. Wallace remarks, the natives in all countries are glad to protect their naked backs and shoulders with some slight covering. **No one supposes that the nakedness of the skin is any direct advantage to man; his body therefore cannot have been divested of hair through natural selection.** *Nor, as shown in a former chapter, have we any evidence that this can be due to the direct action of climate, or that it is the result of correlated development."* [4] [bold mine]

Darwin now acknowledged that in the tropics of Africa, from where he assumed man had originated, it is almost entirely inhabited today by mammals that are covered in fur or thick hair. He wrote:

"The fact, however, that the other members of the order of Primates, to which man belongs, although inhabiting various hot regions, are well clothed with hair, generally thickest on the upper surface on the head of man being covered with long hair; also on the upper surfaces of monkeys and of other mammals being more thickly clothed than the lower surfaces... is opposed to the supposition that man became naked through the action of the sun." [5]

Man's near hairless body was a paradox that Darwin pondered over a great deal. There had to be a solution and this is where sexual selection comes into the picture. Darwin goes on to say:

"The view which seems to me the most probable is that man, or rather primarily woman, became divested of hair for ornamental purposes, *as we shall see under Sexual Selection; and, according to this belief, it is not surprising that man should differ so greatly in hairiness from all other Primates, for characters, gained through sexual selection, often differ to an extraordinary degree in closely related forms."* [6] [bold mine]

This may sound like a lot of mumbo jumbo, but in essence Darwin hinted that there was another theory which describes how the female (some males too) can be attracted by certain characteristics of the opposite sex such as form, colour and behaviour to the extent that modifications could be brought about in the species including hairlessness. It was not simply the case of the survival of the fittest by natural selection that was involved here - something more subtle was also at work behind the scenes.

THE CARD UP DARWIN'S SLEEVE

Darwin could not counter Wallace's arguments at this time because he knew that Wallace was right, at least as far as natural selection was concerned. However, Darwin was engaged in working on a second theory that was to supplement natural selection and it was this theory that he believed would resolve the disagreement that he and Wallace had about Man's naked hairless condition. He had hinted what he had in mind in his famous book on Natural Selection, namely *The Origin of Species*. In the book Darwin wrote:

> "*This leads me to say a few words on what I have called Sexual Selection. This form of selection depends, not on a struggle for existence in relation to other organic beings or to external conditions, but on a struggle between the individuals of one sex, generally the males, for the possession of the other sex. The result is not death to the unsuccessful competitor, but few or no offspring... thus Sir R. Heron has described how a pied peacock was eminently attractive to all his hen birds. I cannot here enter on the necessary details; but... **I can see no good reason to doubt that female birds, by selecting, during thousands of generations, the most melodious or beautiful males according to their standard of beauty, might produce a marked effect.**" [7] [bold mine]

The reader may not grasp at this stage what Darwin meant by his statement about sexual selection with respects to Man's naked condition, but at that time of the publication of *Origins* his thoughts on Sexual Selection were still being developed. When he finally did publish the book that would discuss his new theory he would say that not all traits evolve because they enhance survival as in the case of natural selection. There would be other traits that were selected because they enhanced an individual's ability to attract a mate. It was this process that Darwin would call sexual selection and he would expand upon Sir Robert Heron's observation regarding the peacock in the book. In fact it became the central argument in support of his new theory.

In a letter to William Bernhard Tegetmeier on the 30th March, 1867 Darwin wrote: "*Nevertheless I am still inclined from many facts strongly to believe that the beauty of the male bird determines choice of female with wild Birds, however it may be under domestication. Sir R. Heron has described how one pied Peacock was extra attractive to the Hens. This is a subject which I must take up as soon as my present book is done.*" [8]

William Bernhard Tegetmeier (1816 - 1912) was an English naturalist (with an interest in pigeons, fowl, and bees), and a founder of the Savage Club. He was a writer and journalist of domestic science in the journal *Field* between 1864 and 1907. Sir Robert Heron (1765 - 1854) was not a naturalist but a politician who had inherited his baronetcy and extensive estates in Lincolnshire from his uncle, Sir Richard Heron, 1st Baronet on the latter's death in 1805. As a result Robert owned Stubton Hall, a large estate at Stubton (near Newark-on-Trent) on the border of Lincolnshire and Nottinghamshire and here he kept and bred a large menagerie of animals and birds including llamas, alpacas. lemurs, porcupines, armadillos, kangaroos - and peacocks.

SEXUAL SELECTION IS PUBLISHED

Darwin put in writing his new theory of sexual selection in his second book on evolutionary theory called *The Descent of Man, and Selection in Relation to Sex*. It was published in 1871, twelve years after his famous *Origins* book. *Descent* discussed many related issues, including evolutionary psychology, evolutionary ethics, differences between human races, differences between sexes and the relevance of the evolutionary theory to society. However, as far as Sexual Selection was concerned he emphasised the dominant role that women had in choosing a male in order to mate with him and in this Darwin made reference to the animal kingdom for supportive evidence.

There are three sections to *Descent*. Sections I and III look at the evidence for the development of humans from more primitive creatures and sexual selection in humans. Section II (about half the book) is devoted to sexual selection in everything from insects to mammals. In the introduction to this section Darwin wrote:

> "*We are, however, here concerned only with sexual selection. This depends on the advantage which certain individuals have over others of the same sex and species solely in respect of reproduction. When, as in the cases above mentioned, the two sexes differ in structure in relation to different habits of life, they have no doubt been modified through natural selection, and by inheritance, limited to one and the same sex.*" [8]

What Darwin was about to unleash through *Descent* was the provocative theory that female choice among competing males lead to diverging racial characteristics that could also lead to hair loss. Then he makes an admission in the book. "*From our ignorance on several points, the precise manner in which sexual selections acts is somewhat uncertain.*"

This is most surprising because Darwin immediately says afterwards:

"Nevertheless if those naturalists who already believe in the mutability of species, will read the following chapters, they will, I think, agree with me, that sexual selection has played an important part in the history of the organic world." [9]

Darwin now spends a considerable amount of time in *Descent* trying to explain his theory, even though he has already acknowledged that the precise manner how sexual selection works is uncertain. Not that this matters because a theory is a theory and not fact, except that many naturalists who already accepted natural selection, the same ones whom Darwin now directed to read his new book, also took what he said as Holy Writ. Darwin was preaching to the already converted.

THE TWO WAYS OF SEXUAL SELECTION

According to Darwin sexual selection works in two ways. First there is "Competition". This is where members of the same sex compete (fight each other) for access to the other sex. Darwin called this the **Law of Battle** and this competitive behaviour, he says, may take on many forms. A common example would be direct combat between males. These combats determined hierarchies in which the dominant males established territories and therefore they would gain access to sexually receptive females living in the area. In cases like these, horns, antlers, or other combative devices are acted upon by sexual selection because they are directly involved in helping an individual obtain a mate.

The second way for sexual selection is down to "Female Choice and Ornamentation". Here Darwin argued that while there was direct physical competition between members of the same sex, another factor also enters into the frame. This was how ornamentation was used by the male to grab attention from the opposite sex to attract a perspective mate. In this Darwin pointed to the female of the species as being the decision maker in this regard and he referred to the example of the peacock with its colourful array of tail feathers as being particularly alluring to peahens as evidence behind his theory.

Let us take a look at these two ways of sexual selection in greater detail and see if they can offer a valid explanation for Man's loss of hair, assuming that he had any in the first place.

1. Competition

With regards to **competition** Darwin said in **The Descent of Man:**

> "It is certain that amongst almost all animals there is a struggle between the males for the possession of the female. This fact is so notorious that it would be superfluous to give instances. Hence the females have the opportunity of selecting one out of several males, on the supposition that their mental capacity suffices for the exertion of a choice. In many cases special circumstances tend to make the struggle between the males particularly severe. Thus the males of our migratory birds generally arrive at their places of breeding before the females, so that many males are ready to contend for each female." [10]

There is something wrong with Darwin's statement. First he argues that when the struggle between the males is done the female can then choose the victor to mate with. In fact the opposite is true. It is the victor having established supremacy of his territory who chooses which females to mate with. Females simply do not have a choice.

Secondly, saying that almost all animals compete for the females is also inaccurate. There are numerous examples in the animal kingdom that contradicts his statement. Take the male lion for example. Nearly all reproduction in lions is done by resident prides, which usually consists of about six related females, their dependent offspring, and a "coalition" of 2-3 resident males that joined the pride from elsewhere. Male lions who live together within the same pride do not compete with each other over mating rights with females.

> "Nearly all reproduction in lions is done by the resident prides. So, if you're not in a pride...you're not mating. Mating within a pride is first come first served. Since there is no dominance hierarchy amongst males or females, no mate selection is found to occur." [11]

Likewise, as for as our so called nearest relative the chimpanzee is concerned their troops are primarily composed of related males, hence there is little to no outward signs of competition for females. When a chimpanzee female comes into a state or period of heightened sexual arousal and activity she will mate pretty much non-stop until she comes out of it. Lines will form at the base of a tree in which she dwells, and the chimpanzee males will simply wait their turn, ascend the tree, and rely on their penis and sperm to do all the competitive work. Sexual selection regarding female choice has no bearing on the matter in these examples.

When we look at the first way sexual selection is described by Darwin, the law of battle as he calls it, what we are really seeing is natural selection, pure and simple. Animals fighting each other to establish territories and being able to have the pick of females in these areas is really just natural selection's 'survival of the fittest' acting within the same species.

It does make perfect sense that the strongest animal will prevail in any competition for territory and it will therefore have the choice of available females in the area. That being the case then the winner of the competition will pass on its attributes and traits to the next generation. As any modern dictionary will tell you:

> "*Natural selection is the process by which forms of life having traits that better enable them to adapt to specific environmental pressures, as predators, changes in climate, or **competition for food or mates**, will tend to survive and reproduce in greater numbers than others of their kind, thus ensuring the perpetuation of those favourable traits in succeeding generations.*" [12] [bold mine]

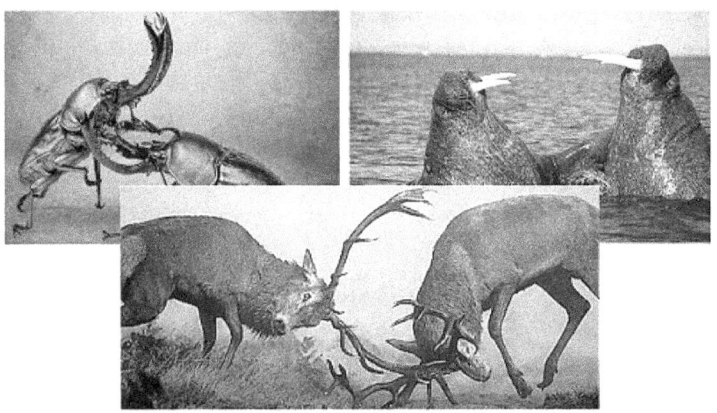

What Darwin was really talking about under the guise of sexual selection is just nature's selective breeding in action. Natural selection replicates artificial selection where horse breeders choose the strong traits of a horse, for example, to breed a thoroughbred that has the strength and stamina to run in races. Incidently, this process whether through man's intervention or natural process is called Microevolution as most dictionaries well tell you.

> "**Microevolution:** *evolutionary change involving the gradual accumulation of mutations leading to new varieties within a species*". (dictionary.com) "*Small-scale evolution consisting of*

genetic changes occurring usually within a single species and over a shorter period of time than in macroevolution." (The Free Dictionary)

I and most if not all Creationists do not object to Microevolution because they see the term meaning **Variation within a Kind** which does not conflict with the Biblical account of Creation. Plants and animals were originally created with large gene pools within distinct created kinds. A large gene pool gives a created kind the genetic potential to produce a variety of types within the kind, allowing the offspring to adapt to varying ecosystems and ensure the survival of that kind of organism.

Returning to the subject of sexual selection, in *Descent* Darwin commented that, *"...sexual selection is, therefore, less rigorous than natural selection"* but then he goes on to say: that, *"Generally, the most vigorous males, those which are best fitted for their places in nature, will leave most progeny."* This is natural selection in a nutshell and nothing more. However, Darwin now adds a twist to the theory that he and Wallace had advocated. He goes on to say:

> *"But in many cases, victory will depend not on general vigour, but on having special weapons, confined to the male sex. A hornless stag or spurless cock would have a poor chance of leaving offspring. Sexual selection by always allowing the victor to breed might surely give indomitable courage, length to the spur, and strength to the wing to strike in the spurred leg, as well as the brutal cock-fighter, who knows well that he can improve his breed by careful selection of the best cocks."* [13]

Darwin hinted to his readers that while the vigour of the combatants is important, equally so are the "weapons" that they bring to bear in the struggle for supremacy. He nonetheless refers to the 'brutal cock-fighter' to support his case by saying that the cock-fighter can improve his breed of cockerel by careful selection of the best cocks.' In other words Darwin is making reference to artificial selection and therefore by association he is also talking again about natural selection while attributing the law of battle to sexual selection.

If truth be known, Darwin presents a very confusing picture indeed. Is it natural selection or sexual selection that he his talking about when describing competition in the context of which we have been discussing? Your guess is as good as mine, but with his second theory of sexual selection Darwin muddies the waters even further.

2. Female Choice and Ornamentation

Darwin speculated that in the past progenitors of the peacock had short tails and was merely spotted with some colour, quite ordinary in fact. He wrote in *Descent:*

> *"As far, then, as gradation throws light on the steps by which the magnificent train of the peacock has been acquired, hardly anything more is needed. If we picture to ourselves a progenitor of the peacock in an almost exactly intermediate condition between the existing peacock, with his enormously elongated tail-coverts, ornamented with single ocelli, and an ordinary gallinaceous bird with short tail-coverts, merely spotted with some colour, we shall see a bird allied to Polyplectron-that is, with tail-coverts, capable of erection and expansion, ornamented with two partially confluent ocelli, and long enough almost to conceal the tail feathers, the latter having already partially lost their ocelli. ... Many female progenitors of the peacock must, during a long line of descent, have appreciated this superiority; for they have unconsciously, by the continued preference for the most beautiful males, rendered the peacock the most splendid of living birds."* [14]

From this ➡ To this

When Darwin spoke of "an ordinary gallinaceous bird" he was referring to a bird relating to or resembling the domestic fowl. Over time peahens showed a preference to mate with males with slightly longer and more flamboyant than average tails. Thus, the characteristic of slightly longer, more brightly coloured tails would be passed on to the next generation and over many generations' peacock tails would become longer and brighter. Thus, said Darwin, the ornate tail gave the peacock an advantage in terms of mating success over its rivals.

Darwin knew that he was treading on shallow water with his explanation. He said:

> "*Many will declare that it is utterly incredible that a female bird should be able to appreciate fine shading and exquisite patterns. It is undoubtedly a marvellous fact that she should possess this almost human degree of taste.*" [15]

Of course there is no evidence to support his hypothesis, which he calls "a marvellous fact", that female peacocks have the same kind of affinity to beauty that humans have, but Darwin continued with his theory regardless. He had no option because there was no other explanation that he could offer that could account for Man's hairlessness. Darwin was not really happy to promote the idea that peahens appreciated beauty like humans, and it made him feel sick at the thought. This is recounted in a letter in 1860 to Asa Grey Darwin.

> "*It is curious that I remember well time when the thought of the eye made me cold all over, but I have got over this stage of the complaint, & now small trifling particulars of structure often make me very uncomfortable. The sight of a feather in a peacock's tail, whenever I gaze at it, makes me sick!*" [16]

Darwin had convinced himself that the answer had to do with female choice and the attractive qualities exhibited by the male. The example of the peacock was his best shot in proving his theory of sexual selection but trying to make a case with this bird's mating ritual was not going to be an easy task.

For Darwin to say that peacocks were coloured like common fowls and having shorter tails than what peacocks have now in the past was complete conjecture on his part because there was no evidence of any kind that this had been the case. In fact, for thousands of years the peacock and its beautiful tail feathers was the object of many myths and legends of ancient Egypt, India, Persia, China, Greece and Rome.

In ancient Greece for example, the peacock was the patron bird of the Goddess Hera. According to myth, she placed 'eyes' on its feathers, symbolising all-seeing knowledge and the wisdom of the heavens. Hence, for thousands of years the peacock had not changed its appearance but Darwin speculated that if one accepted the principle of gradual evolution, then this was not always the case.

> "*If we admit the principle of gradual evolution", Darwin said, "there must formerly have existed many species which presented every successive step between the wonderfully elongated tail-coverts of the peacock and the short tail-coverts of all ordinary birds.*" [17]

Darwin's conjecture that **there must be** transitional forms of birds showing every successive steps between elongated elongated tail-coverts of the peacock and the short tail-coverts of all ordinary birds have not been found in the fossil record to the present day. He was wrong with his assumption.

THE PEACOCK PROOF DISCREDITED

Darwin used the mating ritual of the peacock to attract peahens as his primary proof of his theory of sexual selection. However, his hypothesis had not been proved or disproved through tests using the scientific method. His theory was based upon the observations of how peacocks and peahens behaved during mating, and for all intents and purpose, it would seem that his arguments were sound. It would be a hundred years or more before the scientific method was brought to bear on the subject. In the meantime, Darwin was convinced that he had the answer to the anomalies that natural selection could not explain, including man's hairless body.

While it is true that peacock tail feathers do play a role in the courtship ritual of peafowl this did not necessarily mean that the female was 'attracted' to the feather display. When the peahen observes the feathers of the peacock displayed in all its glory it could be simply a visual cue to say that the peacock was ready for mating. There was no evidence that the peahen has the same understanding of beauty as we humans. She could simply be responding to a peacock's mating ritual through natural instinct. This is confirmed by a recent study in Japan.

Researchers at the University of Tokyo studied peacocks and peahens in a zoo for seven years. They carefully photographed each male during the tail-fanning display ritual and counted the number of eyespots-a measure of the tail's quality. They next examined whether females chose mates with the best-quality tails. During this period of observation, the scientists observed 268 successful matings. But to their surprise, they found that females mated with drab-tailed peacocks as often as with flashy males! They concluded that the male peacocks' tails were not the reason for the females' attraction to a partner - a result at odds with Darwin's theory of sexual selection. [18]

Another study was carried out by Roslyn Dakin, a PhD student in behavioural ecology at Queen's University in Kingston, Canada. She carried out experimental work in 2011 by plucking the feathers of peacocks. She noticed a drop in their success with peahens. However, she also found that, before plucking, males typically had between 165 and 170 eyespots on their trains, and on average, those with the most eyespots didn't mate any more than males with less extravagant tails. Dakin concluded that under most situations, females don't pick mates on the basis of the number of eyespots on their trains, but that the trait could help to weed out particularly unfit males that are missing lots of feathers. [19]

The latest study reported in the August 2013 edition of the *Journal of Experimental Biology, Positive Science* collaborated with evolutionary biologist Jessica Yorzinski to develop eye-trackers small enough for peahens. Scientists from UC Davis and Duke University then rigged twelve birds with the headgear equipped with two cameras - one aimed at one of the bird's eyes, the other at the scene in front of the animal. In the experiments, each peahen was placed alongside two peacocks in an outdoor enclosure. What happened next was surprising.

The male peacocks vied for attention with elaborate shows, shaking their wings and rattling their train of plumes, hooting and dashing toward peahens if the females appeared ready to mate. The scientists weren't surprised when they saw that the peahens looked at each peacock when it fanned its tail feathers upward. But, intriguingly, the females focused nearly completely on the bottom part of the train, close to the ground.

They mostly ignored the conspicuous upper fan. This suggested that the flashy upper train is mostly a long-distance attraction signal, but the lower feathers are more important to close-up courtship. [20]

Darwin had got it wrong! And there was more. There is one other factor to be taken into consideration. Sexual selection, if the theory was true, would no doubt produce features that would reduce the ability for an animal or bird to escape from predators because aesthetic features often made a creature more conspicuous. The colourful fanned tail of the peacock for example would not only make the bird stand out for miles but the weight of the tail feathers would no doubt also slow the bird down, making it easy pickings for such a predator.

Clearly, there is a lot of things wrong with sexual selection and we have only just scraped the surface, but regardless of this, what has this troubled theory got to do with humans becoming hairless?

HAIRLESSNESS AND SEXUAL SELECTION

In *Descent* Darwin begins to explain the enigma of why Man became hairless by pointing to women as has having been denuded first, sometime in the distant past.

> "*The view which seems to me the most probable is that man, or rather primarily woman, became divested of hair for ornamental purposes ... As the body in woman is less hairy than in man, and as this character is common to all races, we may conclude that it was our female semi-human ancestors who were first divested of hair, and that this occurred at an extremely remote period before the several races had diverged from a common stock.*" [21]

Just as the Church blamed Eve for original sin, Darwin pointed the finger at women as being instrumental in Man's hairless condition.

> "*Whilst our female ancestors were gradually acquiring this new character of nudity, they must have transmitted it almost equally to their offspring of both sexes whilst young; so that its transmission, as with the ornaments of many mammals and birds, has not been limited either by sex or age.*" [22]

Darwin uses the words 'must have' because he knew that otherwise there would be no other explanation. So in order to emphasise this point he shrugged his shoulders, metaphorically speaking, and then said:

> "*There is nothing surprising in a partial loss of hair having been esteemed as an ornament by our apelike progenitors, for*

we have seen that innumerable strange characters have been thus esteemed by animals of all kinds, and have consequently been gained through sexual selection." [23]

What Darwin was saying was that our female apelike progenitors - his words by the way - began to lose hair for reasons he does not explain, but the hairless area was so attractive that the females preferred to mate with them rather than their more hairy neighbours. As time went on, the ornamental aspects of loss of hair came into play as each generation of humans shed their hair bit by bit with men preferring the hairless skin of their partner, and likewise women seeking the same in their prospective mates. Sooner or later, millions of years really, natural selection would have culled the hairier males while the hairless ones would have prevailed, survived and conquered. By losing hair bit by bit over a long period Darwin considered this would not be injurious to man.

"Nor is it surprising that a slightly injurious character should have been thus acquired; for we know that this is the case with the plumes of certain birds, and with the horns of certain stags." [24]

What proof is there in what Darwin proposed as the reason for man's hairless condition? There is none. Darwin could only offer some anecdotal evidence, which really had no bearing on the matter.

"The females of some of the anthropoid apes, as stated in a former chapter, are somewhat less hairy on the under surface than the males; and here we have what might have afforded a commencement for the process of denudation." [25]

That was it! This was not evidence. This was conjecture plain and simple and Darwin knew it. So what does he do? He refers to a common proverb that existed at the time.

"With respect to the completion of the process through sexual selection, it is well to bear in mind the New Zealand proverb, 'There is no woman for a hairy man.'" [26]

Darwin, by expressing the aforementioned proverb, was saying that women preferred hairless men. This was hardly proof for his theory. It smacks of an act of desperation to convince people that women preferred hairless men and this was the reason why man evolved to become the hairless being that he had become.

According to Darwin "There is no woman for a hairy man" or by implication
"There is no female for a hairy ape male." This is RUBBISH of course,
but this was considered true and scientific by Darwin at the time.

Darwin should have taken a closer look at the couples that walked the streets of the town where he lived. Outward expressions of hairiness was clear to see in what became known as the 'beard movement'. The reasons for this are more certain. By the mid-nineteenth century, the British Empire was in full flower, and the power of the British military was a matter of pride at home. Some military regiments had begun to wear moustaches, and British men began to imitate this style, with all its attendant military, masculine associations.

There were other factors too. This was also the age of explorers who headed out into untamed lands living amongst wild nature. Such men were the heroes of their day. Often unable to shave 'in the field' they sported large beards, and to imitate this was to link themselves with a symbol of natural virility and masculinity, a visual symbol of innate manliness. [27]

Darwin was clearly blinded by his own prejudices and his perceptions about women. He should have taken a look at himself in the mirror, and realised what he had said was rubbish. By the time of the publication of **Descent** he had lost his head hair and was sporting a beard that was long and untidy, as well as having thick unmanaged eyebrows. Even when he was younger and on the Beagle he had long hair, lengthy bushy sideburns and thick eyebrows. This did not stop him from attracting his future wife Emma or prevent him from being sexually active with her. He had ten children, three of whom died at early ages.

Many recent studies have shown that some women prefer hairless men and others do not. As the on-line magazine **LikeCosmetics** points out.

"In one of the recent studies conducted on sexual behaviour and relationships, it has come out that women prefer hairy hunks to the effeminate metro-sexual men, because they feel a kind of safety and security factor attached to these rough-n-tough men. Their broad muscular bodies and the hairy chests talk about the hard work and the manly things that they do. And that seems like an immediate turn on for most women, no wonder women get the wild vibes looking at sweating mechanics and plumbers in their greasy uniforms, who live outside the peripheries of the overly sophisticated society." [28]

SEXUAL SELECTION IS REJECTED

Although Darwin had put forward his theory of sexual selection to account for Man's hairlessness, his co-discover of natural selection Alfred Wallace rejected it (as did many others). Writing in the same year that Darwin's **Descent of Man** was published in the magazine **The Academy** Wallace said:

*"Natural selection does not produce absolute perfection but only relative perfection...The vast amount of the superiority of man to his nearest allies is what is so difficult to account for. His absolute erectness of posture, **the completeness of his nudity**, the harmonious perfection of his hands, the almost infinite capacities of his brain, constitute a series of correlated advances too great to be accounted for by the struggle for existence of an isolated group of apes in a limited area."*[29]

Wallace argued that the aspects of it which were male to male competition, while real, were simply forms of natural selection, and that the notion of "female choice" was attributing the ability to judge standards of beauty to animals far too cognitively undeveloped to be capable of aesthetic feeling (such as beetles). He also argued that Darwin favoured too much the bright colours of the male peacock as adaptive without realising that the "drab" peahen's colouration is itself adaptive, as camouflage.

For Wallace, writing a few years after Darwin's death, it was clear that Darwin by invoking sexual selection as a mechanism had strayed from the true path. Wallace said in his book **Darwinism** that he published in 1889 that:

> "*Although I maintain, and even enforce, my differences from some of Darwin's views, my whole work tends forcibly to illustrate the overwhelming importance of Natural Selection over all other agencies in the production of new species. I thus take up Darwin's earlier position, from which he somewhat receded in the later editions of his works, on account of criticisms and objections which I have endeavoured to show are unsound. Even in rejecting that phase of sexual selection depending on female choice, I insist on the greater efficacy of natural selection. This is pre-eminently the Darwinian doctrine, and I therefore claim for my book the position of being the advocate of pure Darwinism.*" [30]

Wallace could rightly claim that his theory of natural selection was the purest, as Darwin had allowed other doctrines such as the transmutation of species to taint Darwin's original theory. Natural Selection was natures' equivalent to Artificial Selection, the process by which humans breed other animals and plants for particular traits. But Darwin's original brainchild had become evolution, the combination of his version of natural selection and his grandfather's (and Lamark's) transmutation of species.

Darwin considered sexual selection as integral to evolution as natural selection, but the theory was rejected by most scientific and lay audiences. There were many reasons for this. Predominantly, sexual selection theory clashed with biases prevalent in his day - about culture and nature, mind and body, male and female - in ways that Darwin's first sketch of evolution did not. Whereas *Origins* tracked the development of bodily structures *Descent* did not. It focused not on form but function - on behaviour.

In Darwin's view, the mind was the seat not of reason but of sex. Also, when he postulated that female choice sculpted the male form, he conflicted with the long-standing cultural assumption that men controlled women. And, in discussing female beauty, Darwin showed a fascinating ambivalence about its implications for human gender relations. Finally, he was at odds with his culture in the premium he placed on desire. In short, his theory landed on a cultural fault line, causing first upheaval and mayhem, and then it fell into an abyss that it had helped to create.

Many biologists in Darwin's day and long afterwards rejected the theory of sexual selection. One of these was Thomas Hunt Morgan (1866-1945) an American evolutionary biologist, geneticist and embryologist who became famous for his discoveries relating the role the chromosome plays in heredity. In 1903 he ridiculed Darwin's theory by saying:

> *"Shall we assume that... those females whose taste has soared a little higher than that of the average (a variation of this sort having appeared) select males to correspond, and thus the two continue heaping up the ornaments on one side and the appreciation of the ornaments on the other side? No doubt an interesting fiction should be built up along these lines, but would anyone believe it, and if he did, could he prove it?"* [31]

Since both Wallace and Darwin agreed that natural selection could not on its own account for human hairlessness, and Darwin's second theory of sexual selection also failed to offer a solution too, then how did humankind become hairless? The German pathologist Max Westenhöfer (1871-1957) felt that he had the answer.

> *"The fact that man lacks hair - but probably was hairy at some earlier stage - suggests an analogy with the relative absence of hair in water mammals (whale, sea-cow, hippopotamus), especially since so far there is no other plausible explanation."*

Thus, the theory of the Aquatic Ape was born. It is a theory that is taken seriously by some naturalists as it attempts to answer a lot of unanswered questions that present evolutionary theories find unable to answer, such as our almost hairless condition.

Those who support this theory say that it explains why we lack fur, walk upright, have big brains and subcutaneous fat and have a descended larynx, a feature common among aquatic animals but not apes. They further say that standard evolutionary models suggest these different features appeared at separate times and for different reasons. However, the Aquatic Ape Theory, they say, explains how all of these things could have happened because our ancestors decided to live in or near water for hundreds of thousands or possibly millions of years. But is this theory yet another example of evolutionist desperation to explain our almost naked skin? Let us look at the evidence that is presented by those who uphold this theory and see if this theory can be regarded with any degree of seriousness in the science community.

Chapter 3
THE AQUATIC APE

"The fact that man lacks hair - but probably was hairy at some earlier stage - suggests an analogy with the relative absence of hair in water mammals whale, sea-cow, hippopotamus, especially since so far there is no other plausible explanation."
(Professor Max Westenhofer, "The Road to Man", 1942)

HARDY'S THEORY OF THE AQUATIC APE | THE THEORY TAKES SHAPE | FROM THE ASHES OF THE PHOENIX | THE EVIDENCE OF FAT | OTHER EVIDENCE? | THE THEORY FAILS TO DELIVER

With the failure of Darwin's theory of sexual selection to explain the hairlessness of humans, others sought alternative explanations over the years. With the exception of the head most humans have very little or no body hair. However, it just so happens that whales, dolphins, walruses and manatees also do not have body hair. This has led some researchers to think that if evolution found it advantageous to shed these animal body hairs because they spent more and more time in the water, might not evolution have done the same for humans too. Thus the Aquatic Ape theory (AAH) was born.

It was the German pathologist Max Westenhofer (1871-1957) who first proposed an early version of AAH, which he labelled "the aquatile man", which he described in several publications during the 1930s and 1940's. He disputed Charles Darwin's theory on the kinship between modern man and the great apes.

As part of a complex and unique presentation of human evolution, Westenhofer argued that a number of traits in modern humans derived from a fully aquatic existence in the open seas, and that humans only in recent times returned to land. In 1942, he stated:

> *"The postulation of an aquatic mode of life during an early stage of human evolution is a tenable hypothesis, for which further inquiry may produce additional supporting evidence."* [1]

Westenhofer reasoned that the fact that man lacked hair - but probably was hairy at some earlier stage - suggested an analogy with the relative absence of hair in water mammals (whale, sea-cow, hippopotamus), especially since so far there was no other plausible explanation to date. He further reasoned that the subcutaneous layer of fat in humans; its capacity for expansion appears to predate human civilisation.

As evidence of his theory he pointed to the so-called Venus statuettes, dating back to the Stone Age. He said that "For example, to me the story of Beowulf's struggle with the dragon under water is a hint that humans perhaps did live and fight with such dragons in water." Astonishingly, Edgar Dacqué came to similar conclusions at just about the same time in his interesting and stimulating book *Urwelt, Sage und Menschheit (primeval world, mythology and humanity)* about the coexistence of humans and dragons. [2]

Westenhofer knew that his hypotheses would at first glance appear outrageous, but he argued that it should only serve as initiative in which direction one can research, just as a detective who follows every clue, no matter how unlikely it seems, along with those that seem more promising. Needless to say, Westenhofer's theory fell on deaf ears and that would have been the end of it had it not been for a professor in England who independently came to a similar conclusion. His name was Alister Hardy.

HARDY'S THEORY OF THE AQUATIC APE

Sir Alister Clavering Hardy, (1896 - 1985) was an English marine biologist, expert on zooplankton and marine ecosystems and a highly respected scientist. He was the first Professor of Zoology at the University of Hull from 1928 - 1942, after which he was appointed Professor of Natural History at the University of Aberdeen, where he remained until 1946. He then became Linacre Professor of Zoology in Oxford, a position he held until 1961. In 1940, he was made a Fellow of the Royal Society and was knighted in 1957. Other honours included the Templeton Prize, which he received shortly before his death in 1985. Hence, in view of his reputation as a scientist any theory that came from his pen should be worthy of note.

Hardy was the zoologist on the RRS Discovery voyage to explore the Antarctic between 1925 and 1927, as part of the Discovery Investigations. Through his studies of zooplankton and its relationship with predators, he became expert in marine mammals such as whales. It was in 1930, while reading naturalist and anthropologist, Frederick Wood Jones'*Man's Place among the Mammals*, which included the question of why humans, unlike all other land mammals, had fat attached to their skin, Hardy realised that this trait sounded like the blubber of marine mammals.

From this titbit of information he began to suspect that humans had ancestors that were more aquatic than previously imagined. Fearing a backlash against such a radical idea, he kept this hypothesis secret until 1960, when he openly spoke about for a short time, and later wrote on the subject, which subsequently became known as the aquatic ape hypothesis in academic circles.

In his old age in a filmed interview Hardy said frankly, "*I wanted to be a professor. I wanted to be a Fellow of the Royal Society.*" *There were things you did not do if you hoped for those rewards. In 1960, he had achieved both of those objectives and a knighthood into the bargain, so he slightly relaxed his vigilance and confided his thoughts, just for fun, to a small lay audience - a sub aqua club in Brighton. However, unbeknown to Hardy an enterprising local reporter was present and before long national Sunday newspapers carried banner headlines: "Oxford professor says man a sea ape*". [3]

THE THEORY TAKES SHAPE

When Hardy finally voiced his thoughts about the possibility of an aquatic ape in a speech at the British Sub-Aqua Club in Brighton on 5

March 1960, he found that several national newspapers had distorted the presentation of what he had said. So he countered by explaining them more fully in an article in New Scientist on 17 March 1960, under the heading, **Was there a Homo aquaticus?** [4]

Hardy was very serious about his theory and believed that although it seemed far-fetched it was the best explanation for the striking physical differences that separate man's immediate ancestors from the more apelike forms which had supposedly diverged from a common stock of more primitive tree-living apelike creatures.

WAS MAN MORE AQUATIC IN THE PAST?

And was it in the sea that Man learned to stand erect? The author explains his hypothesis that we descend from more aquatic ape-like ancestors. He will contribute a further article—"Will Man be more aquatic in the future?"—to next week's issue.

by Professor Sir ALISTER HARDY, FRS

In water, only the head needs protec-from the rays of the tropical sun.

Hardy's thesis was that a branch of our apelike ancestors was forced by competition from life in the trees to feed on the sea-shores. They found food such as shell fish, sea-urchins, etc., in the shallow waters off the coast, in the warmer parts of the world. Hardy allowed his imagination to run riot and imagined them wading at first perhaps still crouching, almost on all fours, groping about in the water, digging for shell fish, but becoming gradually more adept at swimming. Then, in time, Hardy said, he could see this creature becoming more and more of an aquatic animal going farther out from the shore; diving for shell fish, prising out worms, burrowing crabs and bivalves from the sands at the bottom of shallow seas, and breaking open sea-urchins, and then, with increasing skill, capturing fish with his hands.

To support his hypothesis, Hardy offered some observational evidence.

> "*First and foremost perhaps, is the exceptional ability of Man to swim, to swim like a frog, and his great endurance at it. The fact that some men can swim the English Channel (albeit with training), indeed that they race across it, indicates to my mind that there must have been a long period of natural selection*

improving Man's qualities for such feats. Many animals can swim at the surface, but few, terrestrial mammals can rival Man in swimming below the surface and gracefully turning this way and that in search of what he may be looking for." [5]

Regarding Man's hair loss, Hardy had an explanation. He said that whilst not invariably so:

"...the loss of hair is a characteristic of a number of aquatic mammals; for example, the whales, the Sirenia (that is, the dugongs and manatees) and the hippopotamus. Aquatic mammals which come out of the water in cold and temperate climates have retained their fur for warmth on land, as have the seals, otters, beavers, etc. Man has lost his hair all except on the head, that part of him sticking out of the water as he swims: such hair is possibly retained as a guard against the rays of the tropical sun, and its loss from the face of the female is, of course, the result of sexual selection." [6]

Hardy further argued that hair, under water, naturally loses its original function of keeping the body warm by acting as a poor heat conductor. He next puts forward some supportive evidence from Dr Weiner, another scientist.

"Man, having lost his hair, must, before he acquired the use of clothing, have been subjected to great contrasts of temperature out of water; in this connection it is interesting to note the experiments carried out at Oxford by Dr. J. S. Weiner, who showed what an exceptional range of temperature change in air Man can stand, compared with other mammals." [7]

Weiner was one of the scientists that uncovered the Piltdown Man Hoax [8] so Hardy calling on him for scientific support, although not for his theory but for science in general, was a good move on his part.

Another observation that seemed to support his theory was that the graceful shape of man or woman! It is most striking when compared with the clumsy form of the ape. All the curves of the human body have the beauty of a well-designed boat. Men and women are indeed truly streamlined.

Then there was the matter of body fat. Hardy said that in 1929 when he returned from an Antarctic expedition he had observed that the layers of blubber of whales, seals and penguins were such a feature of these examples of aquatic life he wondered if Man too might have been so equipped. He explained that warm-blooded water animals had such layers of fat which act as insulating layers to prevent heat loss; in fact, in function they replace the hair. Hardy pursued this argument by describing Man's great number of sweat glands that enabled him to stand a tropical climate while still retaining a large layer of fat necessary for aquatic life.

Hardy's idea received some interest after the article in the magazine *New Scientist* in 1960 was published, but it was generally ignored by the scientific community. As for the authorities at Oxford University where he lectured they were livid. He had not followed protocol and consulted his colleagues, nor had he submitted his ideas to peer review. He was not and never had been an anthropologist, and the authorities said that he had exposed Oxford science to public ridicule by publicly presenting such a bizarre and childish notion.

Hardy saw the error of his ways. He was permitted give one talk on the BBC's *Third Programme*, to correct some garbled versions of what he had actually said, but that was to be the end of it. Sure enough, he never made mention of it again and after his death, when a memorial service was held to honour his memory, no mention was made of this incident in his life. It was as if it had been a discreditable aberration on his part which it was kinder to forget.

In 1967, Hardy's hypothesis was briefly mentioned in *The Naked Ape*, a book by Desmond Morris in which can be found the first use of the term "aquatic ape", but beyond this fleeting mention the theory seemed to have died a natural death and buried - until science writer Elaine Morgan OBE FRSL 1920-2013) entered the fray.

Elaine Morgan's credentials speak for themselves. She had won two BAFTAs and two Writers' Guild awards with her work. She was awarded an honorary Doctor of Letters (equal to the Doctor of Science) by Glamorgan University in December 2006, an honorary fellow of the University of Cardiff in 2007, and awarded the Letten F. Saugstad Prize for her "contribution to scientific knowledge". She was appointed Officer of the Order of the British Empire (OBE) in the 2009 Birthday Honours for services to literature and to education. She was elected a Fellow of the Royal Society of Literature the same year.

FROM THE ASHES OF THE PHOENIX

Elaine Morgan was an avid believer in the aquatic ape hypothesis and she wrote a number of books on the subject that included: *The Descent of Woman (1972), The Aquatic Ape (1982), The Aquatic Ape Hypothesis (1997)*, and *The Naked Darwinist* (2008), which discusses the reception of aquatic scenarios in academic literature.

Morgan said that she had not heard about Hardy's theory until she saw a reference to it in Desmond Morris's *The Naked Ape*. She said that *"Desmond, who knew and admired Alister, gave a brief account of the theory, but concluded that if there had been an aquatic interlude, its effects were probably minimal."* [9]

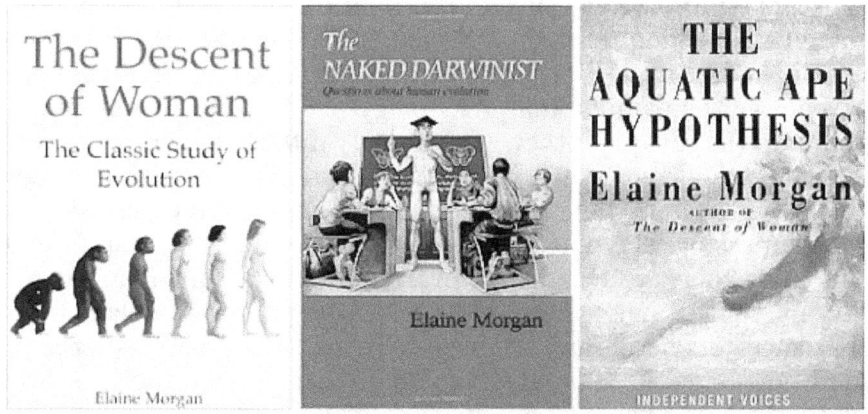

Intrigued by what Hardy had suggested, Morgan wrote to, now Sir Alister Hardy, for permission to follow up the suggestion he made, and he agreed. The result was *The Descent of Woman* and it became a best seller in the USA and book of the month. In the book she mentioned the unmentionable - the penis. On this Morgan wrote:

"The penis thing was relatively very minor. There had been a dispute over why the penis in humans is so much larger than in other apes. Some claimed it was in order to allure females, others that it was to impress and intimidate other males. I made the purely utilitarian observation that bipedalism had made the vagina relatively inaccessible, to the point where a gorilla-sized penis would be totally useless. This was accepted, with the proviso that it would have happened for whatever reason the species became bipedal, so it could not - repeat not - be advanced as an argument for the Aquatic Theory. Fair enough." [10]

Morgan was expecting a backlash from the scientific community, but instead she was simply ignored. For twenty years her book remained in the doldrums. Nobody was interested in what it had to say, simply because, as I personally have found, publishers are not interested in new theories, especially when unsupported by well-known scientists. And as Morgan relates, "there was the intellectual high ground occupied by the PAs - the palaeoanthropologists. They alone knew, as the saying goes, where the bodies were buried. They knew that if you wanted to find fossils of prehuman creatures you had to go out onto the savannah to look for them. And lo! There they were." [11]

Ten years later Morgan tried again. She published her book **The Aquatic Ape**, which from the title made it clear what the book was all about. Expanding upon Hardy's theory and observations in greater detail, she began by explaining that AAT suggests that when our ancestors moved onto the savannah they were already different from the apes; that nakedness, bipedalism, and other modifications had begun to evolve much earlier, when the ape and human lines first diverged from a common ancestor. She pointed out that most of the "enigmatic" features of human physiology, though rare or even unique among land mammals, are common in aquatic ones. Thus, she said, if we postulate that our earliest ancestors had found themselves living for a prolonged period in a flooded, semi-aquatic habitat, most of the unsolved problems become much easier to unravel.

Morgan emphasised one of the greatest mysteries that palaeontologists have failed to answer and that was why Man is naked and virtually hairless. She wrote:

"It must have been around then that the question 'Why naked?' began to be dropped from the agenda by common consent. There was no conscious decision and no concealed motive. Any

scientist worth his salt concentrates on questions he has some hope of answering, and particularly on those where new and relevant evidence has come to light. Newly discovered fossils of skulls and teeth and femurs and foot bones could be examined, measured, and written up. Compared to that it would be a frivolous waste of time to dream up scenarios about loss of body hair when hair does not fossilise." [12]

Morgan was right of course because to this day the loss of hair in humans continues to perplex palaeontologists and warrant explanation. There certainly have been some weird and futile explanations dreamed up over the years, including dare I say it Morgan's aquatic ape theory. However, let us be magnanimous and allow her to present her case.

Morgan explains that although humans are classed anatomically among the primates, the order of which includes apes, monkeys and lemurs, only humans are naked. Here she is stating what everyone knows. Now what? In her books she questions the traditional evolutionary model that Man had evolved in Africa by moving out of the forests onto the grasslands of the open savannah. The distinctly human features are therefore supposed to be adaptations to a savannah environment, or so palaeontologists say. However, this is a problematic explanation, Morgan counters, because other animals that spend a lot of time in the sun kept their fur for protection from it - including chimpanzees and bonobos. She says that we would expect to find at least some of these adaptations to be parallelled in other savannah mammals. But there is not a single instance of this, Morgan argues, not even among species like baboons and vervets, which are descended from forest - dwelling ancestors.

Having raised doubts about the African savannah scenario, Morgan offers instead an alternative by suggesting that our common ancestors inhabited watery habitats. Accordingly, when our ancestors moved onto the savannah they were already different from the apes; that nakedness, bipedalism, and other modifications had begun to evolve much earlier, when the ape and human lines first diverged.

To support her hypothesis she points to the fact that there are two kinds of habitat known to give rise to naked mammals - a subterranean one or a wet one. Morgan offers the example of the naked Somalian mole rat which never ventures above ground. All other non-human mammals, she says, which have lost all or most of their fur are either swimmers like whales and dolphins and walruses and manatees, or wallowers like hippopotamuses and pigs and tapirs. As for the rhinoceros and the elephant, Morgan suggests that though found on land since Africa became drier, these animals bear traces of a more watery past and seize every opportunity of wallowing in mud or water. [13]

Morgan proceeded to cast further doubts about the standard evolutionary model. By questioning some of the theories that evolutionists have offered to explain humans becoming hairless. She says that it has been suggested by some palaeontologists that humans became hairless, "to prevent overheating in the savannah", but no other mammal has ever resorted to this strategy she says. As Morgan rightly says, a covering of hair acts as a defence against the heat of the sun: that is why even the desert-dwelling camel retains its fur. [14]

Another explanation by palaeontologists that Morgan highlights is, "to facilitate sweat-cooling". Here Morgan is in fine form because what she does now to provide valid arguments against this evolutionary model. She says that:

> "...many species resort to sweat-cooling quite effectively without needing to lose their hair. Hence, there is no known reason why an ape should suffer more from overheating than the savannah baboon. And, especially for a savannah primate, there would be a high price to pay for hairlessness. Primate infants are carried around clinging to their mothers' fur; the females would be severely hampered in their foraging when that no longer became possible." [15]

I will discuss the evolutionary theory of sweat-cooling (thermoregulation) in the next chapter and show that in this Morgan is right by what she says, but that does not make her theory is any more valid.

So far Morgan has not yet presented a viable case other than to confirm that man is hairless for reasons that have not yet been explained by standard evolutionary models which have any credibility. The rest is pure speculation on her part. Morgan may talk about the Somalian mole rat as being hairless and offer this titbit of information to prop up her idea, but then there are many rodents who live in the same area who, although burrowing and living underground, are covered in thick hair. And to say that the rhinoceros and the elephant seize every opportunity of wallowing in mud or water, does not mean that this hints at traces of a watery past, any more than man building and flying in aeroplanes hearkened to a time when he might have had wings.

THE EVIDENCE OF FAT

Morgan now turns to biology and what she considers the key evidence of our aquatic past. She says that one general conclusion seems undeniable from an overall survey of mammalian species: that while a coat of fur provides the best insulation for land mammals the best insulation in water is not fur, but a layer of fat. In these humans are by far the fattest primates she says, and we have ten times as many fat cells in our bodies as would be expected in an animal of our size.

So what you may ask? Morgan explains that there are two kinds of animals which tend to acquire large deposits of fat - hibernating ones and aquatic ones. In hibernating mammals the fat is seasonal; in most aquatic ones, as in humans it is present all the year round. Also, in land mammals fat tends to be stored internally, especially around the kidneys and intestines; in aquatic mammals and in humans a higher proportion is deposited under the skin. This would suggest that aquatic mammals and humans are related in some way. She explains:

> "*It is unlikely that early man would have evolved this feature after moving to the plains and becoming a hunter, because it would have slowed him down. No land based predator can afford to get fat. Our tendency to put on fat is likelier to be an inheritance from an earlier aquatic phase of our evolution. It is true that some apes, especially in captivity, may put on weight, but we still differ from them in two important ways. One is that they are never born fat. All infant primates except our own are slender; their lives may depend on their ability to cling to their mothers and support their whole weight with their fingers. Our own babies accumulate fat even before birth and continue to grow fatter for several months afterwards. Some of this fat is white fat, and that is extremely rare in new-born mammals. White fat is not much good for supplying instant heat and energy. It is good for insulation in water, and for giving buoyancy.*" [16]

This all sounds very plausible, and to emphasise her case even further Morgan tells us that when an anatomist skins a cat or rabbit or chimpanzee, any superficial fat deposits remain attached to the underlying tissues. In the case of humans, the fat comes away with the skin, just as it does in aquatic species like dolphins, seals, hippos and manatees. However, her 'evidence' is flawed. While it is true that humans do have subcutaneous fat, it is not different from what the other great apes have, and more to the point it is very unlike the blubbery fat developed by the furless aquatic mammals.

In addition when we look at the life histories of dolphins and whales and other aquatic mammals they rapidly gain fat while very young, and at a very early age they are essentially like their parents in their fat distribution and quantity. In contrast humans start off fairly fat, drop within a few years to the leanest condition of our lives as children, and then rapidly build up fat at puberty, with radical differences in quantity and distribution of fat between boys and girls. And to top it off, at middle age our fat distribution changes once again. All this is very unlike those aquatic mammals whose fat have developed in such a way as to help them deal with that environment.

Then there is the case of the polar bear. The polar bear is well enough adapted for aquatic life and is able to swim easily at enormous distances from land, diving and catching food underwater. They also show physical adaptation to aquatic life in their relatively small, streamlined head compared to their relatives and also absent in humans. If we were to accept what Morgan says that fat differences between such animals are due to living in an aquatic environment either full or part-time, one would expect polar bears' fat quantity and distribution to show differences from their terrestrial relatives. It doesn't. This is just one of many scientific tests of the AAT which it fails. [17]

To support her case Morgan calls upon Caroline Pond, the leading authority on fat and its evolutionary significance in humans and other animals. However, Pond is completely misrepresented by what Morgan says. Morgan's main argument taken from Pond's observations is that humans are on the fat end of the scale compared to other mammals, but conveniently fails to divulge that Pond makes it clear in her writings that the amount of fat in humans is similar to that of captive monkeys if they aren't kept on a strict diet.

Pond responded to the misrepresentations of her work in an article published in **New Scientist** under the heading **Not an aquatic ape -- just an exceptionally fat mammal**. Referring to Morgan and other pro-aquatic ape proponents who have also referred to her work in their literature, Pond takes them to task by saying:

> *"They think that the hair, skin and superficial adipose tissue of humans evolved into "blubber", similar in function to that of seals and whales. Let's consider the facts about the quantity and arrangement of adipose tissue in some aquatic mammals. In specialised aquatic mammals, such as whales, seals and manatees, the limbs are reduced or absent and the trunk is smooth and tapered. But whales and seals are not always exceptionally fat. Massively thick limbs or a bulging abdomen would spoil the streamlining. Adipose tissue around the guts and kidneys is greatly reduced. In all but the most emaciated terrestrial mammals, the mesentery that holds the gut in place is clouded with adipocytes. But in seals, even those that are 50 per cent fat, there is so little adipose tissue in the mesentery that you can read a newspaper through it. The superficial adipose tissue spreads over the trunk forming a continuous layer of blubber that may facilitate rapid swimming by acting as a shock absorber in turbulent water."* [18]

Pond goes on to say:

> *"Less specialised aquatic mammals, such as otters, have elongated bodies and webbed feet, and but the limbs are relatively long limbs [sic] and they can travel some distance over land. The distribution of adipose tissue in otters is almost identical to that of their terrestrial relatives. **No one can claim that the limbs and trunk of humans have evolved further towards fully aquatic habits than those of the otter.** Why should humans have adipose tissue like that of a highly specialised aquatic mammal? Anyhow, as most of us know only too well, there is plenty of fat inside the human abdomen."* [19] [bold mine]

In 1987 a conference, organised by the European Sociobiological Society and the Dutch Association of Physical Anthropology, was held in August 1987. Its aim was to evaluate the pros and cons of Sir Alister Hardy's daring idea about the Aquatic Ape, a presumed early ancestor of humans.

Twenty-two participants went to the Dutch town of Valkenburg to try to come up with some kind of an official statement about the theory. The supporters of the AAH were looking for some kind of vindication of the theory, the opponents were looking for some kind of final rebuttal.

Caroline Pond was one of the most articulate authors arguing against the AAH while Elaine Morgan argued in support of the theory. The book *Proceedings from the Valkenburg Conference* that was written about the proceedings and published in 1991 by Souvenir Press is regarded by all participants as the only serious academic investigation into the plausibility of the so-called "aquatic ape hypothesis". [20]

The conclusion of the conference was somewhat ambiguous but tended to favour against the Aquatic Ape Theory preferring instead to stick to the traditional savannah based model of the palaeontologists. Writing in the epilogue the four editors of the book wrote:

> "*First, it is clearly impossible to provide a conclusive answer to the question of whether there was an aquatic ape. Second the arguments for and against the theory are difficult to weigh against each other. We will not rehearse; they are summarised at the start of each chapter. Our general conclusion is that, while there are a number of arguments favouring the AAT, they are not sufficiently convincing to counteract the arguments against it.*" [21]

OTHER AAT EVIDENCE?

There are three further arguments that AAT proponents offer as evidence. First, we humans are bipedal, that is to say we stand upright on two legs. It is suggested therefore that in water, this made it possible to wade in a greater depth, and for swimming it allowed a coordinated motion of arm strokes and leg kicks as opposed to a clumsy dog paddle.

AAT proponents argue that if man had evolved according to the standard evolutionary model in the savannah, why is it that other savannah animals have not adopted swimming abilities as have humans? From predators like the lion to prey like the antelope, four legs is best for life on the savannah, not two.

This argument is absurd of course. It is a fact that bipedalism has developed only in land animals, and is not an adaptation for an aquatic life. All the mammals who use two legs some or all of the time - kangaroos, primates, bears - are land animals. All the aquatic mammals are either four-legged like hippos, or specialised swimmers like dolphins that use no legs at all.

Secondly, it is noted that we humans can control our breathing consciously, unlike virtually all other animals whose breathing is autonomic. What other mammals are able to do this AAT proponents ask? You guessed it, those who dive. They take a large breath to dive deep, or a shallow breath when swimming casually. However, this argument is flawed too. It is untrue that only humans and aquatic mammals can control their breath voluntarily. Most primates can hold their breath, as can dogs.

While it is true that humans do have much better breath control than any other animal, we also use our breath for speech and other skills not found in the animal kingdom, dolphins included.

Third and finally, humans have sebaceous glands to make our skin oily. It is argued that oily skin is useless on the savannah, but it's quite good for waterproofing, which is the only known use of sebaceous glands in mammals.

In this the AAT proponents do have a point except from one small but relevant detail. While it is true that us humans do indeed have really big sebaceous glands that make our skin nice and oily, we are by no means unique. Another mammal shares this attribute and it is not aquatic. It is the lemur. Why lemurs and humans have sebaceous glands is not thoroughly understood, but there's clearly no correlation between these enlarged glands and swimming.

Oily skin happens when the sebaceous glands produce too much sebum

THE THEORY SIMPLY FAILS TO DELIVER

The Aquatic Ape Theory is an interesting hypothesis that tries to explain why humans are naked and almost hairless. But it falls short in many ways as we have seen. It is a theory that is very difficult to prove or disprove but some noted people have been convinced of its worth. For example, recently, in 2005 Sir David Attenborough produced a television documentary called *Scars of Evolution* a two part series looking at the history and current status of the 'aquatic ape hypothesis' theory. After this it appeared that Sir David had become a convert.

At a conference in 2013 entitled *Human Evolution Past, Present and Future - Anthropological, Medical and Nutritional Considerations* Sir David told reporters that "*the aquatic ape hypothesis made more sense than the conventional savannah narrative accepted by the scientific mainstream. He said that there were serious problems with the conventional hypothesis, while the aquatic theory provided further incentives to preserve marine life. Gathering molluscs is far easier than chasing elephants and wildebeests across the savannah,"* he said. [22]

Professor Chris Stringer of the Natural History Museum, London responds to AAT arguments by saying:

> *"But the whole aquatic ape package includes attributes that appeared at very different times in our evolution. If they were all the result of our lives in watery environments, we would have to have spent millions of years there and there is no evidence for this - not to mention like crocodiles and other creatures would have made the water a very dangerous place."*
> *And, Joe Parker, an evolutionary scientist at Queen Mary University of London, raises the point that: "If our transition to an aquatic or semi-aquatic environment was so successful, why aren't we still there? For the aquatic ape to work, you have to postulate not one, but in fact two unlikely shifts - into an aquatic niche and then back on to land again."* [23]

Morgan's version of the AAH has achieved much popular appeal, but it has never achieved significant acceptance or serious scrutiny within the evolutionist community. John H. Langdon of the Department of Biology, University of Indianapolis in his article published in the *Journal of Human Evolution* summarises the general attitude towards Morgan's theory in evolutionist circles.

> *"The hypothesis is troubled by inconsistencies and has not been reconciled with the fossil record. More importantly, its claim to parsimony is false. The numerous "explanations" for individual anatomical traits that it generates constitute premises that are not better founded than competing terrestrial "explanations".*
> *"* [24]

As both sides of the evolution community battle over whether or not there was such a thing as an aquatic ape, it seems that they have missed the plot. It is one thing for AAT theorists to offer superficial possible indications of an aquatic existence at one time - our lack of hair, our sweat glands, our oily skin and underlying fat, our ability to swim and hold our breath under water, our streamlined shape - but it is a completely different matter to describe a mechanism to explain how we became what we are today with our almost hairless condition. Just by saying that dolphins do not have hair does not mean that man, if one believes in evolution, followed a similar path and descended form an aquatic ape.

While skeletons of ancient dolphins have been found that are reported to be millions of years old, this cannot be said for extinct apes exhibiting any aquatic characteristics. This is not surprising because the reader should keep in mind that AAT theory is part of a mindset, albeit a divergent one to the standard model, that man has evolved. In this the evolution of man demands common ancestors to support the theory.

However, AAT proponents like their evolutionist opponents have not dared to take into account man's incredible penis biology which, I present in the sequel to the present work entitled, *Richard Dawkins Miracle.* This book conclusively proves without a shadow of doubt that there could not have been any common ancestors. As a result, the theory of the evolution of man collapses into a pile of dust, and along with it so does the Aquatic Ape Theory.

Chapter 4
NOW IT GETS HOT!

"There have been a lot of hypotheses made about why we lost most of our body hair. And I definitely, and many colleagues of mine definitely are of the opinion - based on the environmental, anatomical and genetic evidence at hand - that we lost most of our body hair because of the needs of heat regulation."
(Nina Jablonski)

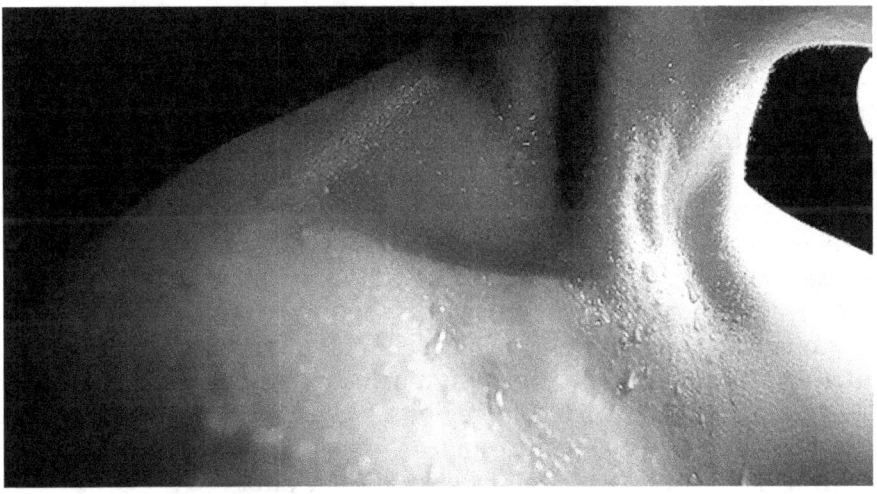

THERMOREGULATION IN HUMANS | PETER WHEELER'S THEORY | FALSE PREMISES | MORE PROBLEMS

Both Darwin and Wallace agreed that natural selection could not account for Man's naked and almost hairless condition and as we have found out neither could sexual selection or the aquatic ape theory offer a plausible solution either. We are left therefore with very few options. However, there is one evolutionary theory that has been paraded around since the 1980s which some evolutionists believe may be responsible for human nakedness and hair loss and that is a hypothesis based around thermoregulation.

Thermoregulation is the ability of an organism to keep its body temperature within certain boundaries, even when the surrounding temperature is very different. What this has to do with Man's hairlessness is difficult to imagine but evolutionists believe they have the answer. Interestingly, evolutionists often castigate creationists for believing in miracles, but they themselves often create miracles to explain and support their theories, and thermoregulation theory for Man's hairlessness is no exception as we shall see.

To regulate body temperature, an organism living in hot arid environments needs to have some mechanism to avoid out of control overheating otherwise death will quickly follow. Evaporation of water is one way to help cool body temperature to within the organism's tolerance range, either across respiratory surfaces or across the skin in those animals possessing sweat glands. In contrast, many animals with a body covered by fur have limited ability to sweat therefore they rely heavily on panting to increase evaporation of water across the moist surfaces of the lungs and the tongue and mouth. Mammals like cats, dogs and pigs, rely on panting or other means for thermal regulation and have sweat glands only in foot pads and snout. The sweat produced on pads of paws and on palms and soles mostly serves to increase friction and enhance grip.

Birds also avoid overheating by gular fluttering, flapping the wings near the gular (throat) skin, similar to panting in mammals, since their thin skin has no sweat glands. Down feathers trap warm air acting as excellent insulators just as hair in mammals acts as a good insulator. Mammalian skin is much thicker than that of birds and often has a continuous layer of insulating fat beneath the dermis. In marine mammals, such as whales, or animals which live in very cold regions, such as the polar bears, this is called blubber. Dense coats found in desert endotherms also aid in preventing heat gain such as in the case of the camels.

THERMOREGULATION IN HUMANS

Humans have been able to adapt to a great diversity of climates through four avenues of heat loss: convection, conduction, radiation, and evaporation and it is his skin that provides the mechanism for regulation. If skin temperature is greater than that of the surroundings, the body can lose heat by radiation and conduction. But if the temperature of the surroundings is greater than that of the skin, the body actually gains heat by radiation and conduction. In such conditions, the only means by which the body can rid itself of heat is by evaporation through sweating.

To facilitate sweating there are 2.6 million sweat glands covering the skin of the human body, everywhere except on our lips, nipples and genitals. Hence, under such circumstances the maximum exposure of as much of our skin to the environment is extremely important to aid the evaporation process. So when the surrounding temperature is higher than the skin temperature, anything that prevents adequate evaporation will cause the internal body temperature to rise. Certain types of clothing therefore can have a detrimental effect on our thermoregulation processes.

There are two kinds of sweat glands in the human body. Eccrine sweat glands cover the entire body just under the skin and these secrete sweat (a fluid containing mostly water with some dissolved ions) which travels up the sweat ducts, through the sweat pores and onto the surface of the skin. This causes heat loss via evaporative cooling; however, a lot of essential water is lost.

Failure to drink sufficient quantities of water to compensate for the loss of water through evaporation will lead to dehydration and ultimately death. Then there are Apocrine glands. These emit a milky substance that regulates perspiration, and they are found mostly under arms and around genitals. Sweat in itself doesn't actually smell. Body odour is caused by the waste products of bacteria that are found naturally on the skin, which thrive in humid sweaty environments, hence odour being more noticeable under the arms.

In general, humans appear physiologically well adapted to hot dry conditions. However, effective thermoregulation is reduced in hot, humid environments such as tropical environments, the Red Sea and Persian Gulf (where moderately hot summer temperatures are accompanied by unusually high vapour pressures), and deep mines where the atmosphere can be water-saturated.

In hot-humid conditions, clothing can impede efficient evaporation. In such environments, it helps to wear light clothing such as cotton, that is pervious to sweat but impervious to radiant heat from the sun. This minimises the gaining of radiant heat, while allowing as much evaporation to occur as the environment will allow. Clothing made from man-made plastic fabrics are impermeable to sweat and thus do not facilitate heat loss through evaporation and will most certainly contribute to heat stress.

In cold conditions sweat stops being produced. The minute muscles under the surface of the skin attached to an individual hair follicle contract, lifting the hair follicle upright. This makes the hairs stand on end which acts as an insulating layer, trapping heat. This is what also causes goose bumps. Humans have very little hair and so goose bumps can easily be seen. Arterioles carrying blood to superficial capillaries under the surface of the skin shrink (constrict), thereby rerouting blood away from the skin and towards the warmer core of the body. This prevents blood from losing heat to the surroundings and also prevents the core temperature dropping further. This process is called vasoconstriction. It is impossible to prevent all heat loss from the blood, only to reduce it. In extremely cold conditions excessive vasoconstriction leads to numbness and pale skin. Frostbite occurs when water within the cells begins to freeze, which destroys the cell causing damage.

Muscles can also receive messages from the thermo-regulatory centre of the brain (the hypothalamus) to cause shivering. This increases heat production as respiration is an exothermic reaction in muscle cells. Also, mitochondria can convert fat directly into heat energy, increasing the temperature of all cells in the body. Brown fat is specialised for this purpose, and is abundant in newborns and animals that hibernate.

This then is a brief overview of thermoregulation in humans. What has this got to do with Man becoming hairless having supposedly evolved from a hairy common ancestor? Editor in Chief Mariette DiChristina of Scientific American explains:

"Our forebears abandoned their easier foraging habits, travelling longer distances through a tropical landscape to acquire sufficient food to survive. Adding meat to their diets meant more calories, but finding prey also took more work. Their activity level increased and with it their need to dissipate body heat to avoid tissue damage. By 1.6 mya, protohumans had long legs for sustained walking and running. Along with that trait came naked skin and a large number of eccrine sweat glands, which produce moisture that removes body heat through evaporative cooling. The hairs on our head also help to combat overheating, by shielding our big brain from direct sun." [1]

Although DiChristina is speaking as if what she says is fact the truth of the matter she is speculating - pure and simple. There is no evidence that the protohumans as she calls them had lost any of their hair during their supposed evolution. Nor is there any evidence that tells us that these hominids abandoned their easier foraging habits to travelling longer distances through a tropical landscape to acquire sufficient food to survive. Are we to say that all the animals including primates living in the area found food scarce?

What is it that has led DiChristina and others to propose that thermoregulation accounts for Man's naked skin and hair loss? Professors Ruxton and Wilkinson of the Institute of Biodiversity, Animal Health and Comparative Medicine, College of Medical, Veterinary and Life Sciences, University of Glasgow provides the answer. In a study they wrote that, *"Thermoregulation is often cited as a potentially important influence on the evolution of hominins, thanks to a highly influential series of papers in the Journal of Human Evolution in the 1980s and 1990s by Peter Wheeler."* [2]

PETER WHEELER'S THEORY

Dr. Peter Wheeler, is professor of evolutionary biology at Liverpool John Moores University, Liverpool, and since 1984 he has published a series of papers on the thermoregulatory advantages of hominin bipedalism combined with naked skin and a larger body size. His theory has gradually gained popularity among evolutionists as being the only viable explanation for Man's naked and almost hairless condition. The key element of his theory is that of bipedalism where an erect posture, he says, exposes less surface area of the body to the sun and lifts the body above the ground with increased exposure to cooling winds, which keeps the body temperature cooler and reduces the dangers of dehydration and sunstroke on the savannah.

Wheeler's first paper **The Evolution of Bipedality and Loss of Functional Body Hair in Hominoids** published in 1984 suggested that an upright posture provides distinct thermoregulatory advantages for a biped over a quadruped in open, equatorial habitats. Specifically, he argued, they would be subject to significantly less direct solar radiation whilst standing in an upright posture at around noon (where the amount of body surface exposed to the sun may be as low as 7%) and also that their upper bodies, being placed higher up away from the ground, would be subject to stronger convective air currents helping them to keep cool though evaporative sweat cooling.

Most of Wheeler's conclusions were drawn from measurements of body profiles in the frontal and vertical planes and estimating similar body profiles in Australopithecines, an extinct genus of hominids that was said to have evolved in eastern Africa around four million years ago and who it was said lived in an equatorial grassland environment.

Most savannah adapted mammals, Wheeler tells us, tend to have an advanced set of features, including a counter current heat exchanging structure known as the carotid rete, a configuration of blood vessels in the brain that can keep its temperature lower than body temperature. As humans and apes lack these features, some other mechanism for cooling the brain would be required if our ancestors had also evolved there. Wheeler's hypothesis is that the upright posture offered by bipedalism may have been a significant part of another such mechanism.

Using two small scale models of a putative hominid ancestors, one positioned in a quadrupedal posture and the other bipedal, Wheeler photographed the models from a range of angles corresponding to position of the sun. His results showed that the bipedal posture significantly reduced the amount of body surface exposed to the sun's rays, particularly around noon.

To analyse whether or not Wheeler's hypothesis has any merit we need to glance into the scientific atmosphere at the time when he proposed is theory. In the 1980's Australopithecus afarensis (better known as Lucy) was considered the oldest known hominin fossil and it was believed that she lived in an equatorial savannah (grassland) habitat as opposed to forested regions. More importantly, it was believed that Lucy walked upright. Australopithecus afarensis therefore gained "human ancestor" status based upon the assumption that for apes to evolve into man, they would have had to walk first. It was from this starting point that Wheeler presented his hypothesis.

FALSE PREMISES

Evolutionist Ken Harding writing on his website in an article headed **You Figure it Out** describes Lucy in comparison to a chimpanzee in the following way.

> "*The Afarensis has a brain of about 415 cc, slightly larger than the chimp. They are obviously both primates, and closely related. But the biggest difference, the difference that absolutely excludes Afarensis, or Lucy (as she is most commonly known) from being a version of chimpanzee, is that she walked upright on two legs (fully bipedal) with a modern looking knee. The chimp has no such ability, and can only sustain partial bipedal locomotion for a few seconds at a time, due to a knee that cannot straighten out. Chimpanzees are not bipedal - they are quadrupeds. Lucy's kind walked the earth 3 to 4 million years ago.*" [3]

Ken agrees that both the skulls of Lucy and chimpanzees resembled each and were closely related. However, his main criteria, obviously gleaned from evolutionary sources, is that the difference that absolutely excludes

Lucy from being a version of chimpanzee, is that she walked upright on two legs. Likewise, talking about Lucy in 2010 the **Smithsonian Magazine** made a similar statement.

> *"At 3.2 million years old, Lucy was remarkably primitive, with a brain and body about the size of a chimpanzee's. But her ankle, knee and pelvis showed that she walked upright like us. This meant Lucy was a hominid - only humans and our close relatives in the human family habitually walk upright on the ground."* [4]

Not all palaeontologists agreed with this assessment. Adrienne Zihlman and Vincent Sarich published a scathing scientific criticism against the idea that Lucy had been an ancestor of man because it was believed that the extinct primate walked upright.

In their article Zihlman and Sarich pointed out the curved finger bones that suggested life swinging in the trees, not standing upright scanning the African plains. They pointed out numerous chimpanzee-like features in Lucy, and even showed a picture of Lucy's skeleton superimposed on the living rain forest or pigmy chimpanzee, Pan paniscus, also called bonobo. A troop of these chimps lives in a fabulous habitat in the San Diego Zoo, and they can occasionally be seen walking upright, just as Lucy ever could. But they are obviously chimpanzees, and not ape-to-man links. No one would become famous, of course, by claiming to find fossil chimpanzees bones in Africa - and Lucy appears to be nothing more." [5]

In 1994 Christine Berge of the Museum of Palaeontology in Paris did a study comparing the pelvis and lower limb of Australopithecus afarensis to that of modern humans, in an attempt to define the pattern of australopithecine bipedal locomotion. Published in the **Journal of Human Evolution** Berge concluded:

> *"Only the reconstruction of the gluteal musculature on the basis of the pongid pattern is consistent with the bony, structure of the fossil and would have permitted effective movements of bipedalism. Moreover, the results clearly indicate that australopithecine bipedalism differs from that of humans. (1) The extended lower limb of australopithecines would have lacked stabilisation during walking; and (2) the lower limb would have shown a greater freedom for motion, which can be interpreted as the retention of a partly arboreal behaviour."* [6]

What Berge is saying is that while Lucy may have been able to walk upright, the extended lower limb would have prevented Lucy from walking like humans. In other words, Lucy was not fully bipedal, and as others have noted would have probably resembled bonobos in their locomotion. Wheeler had assumed that Lucy upon which his assessment had been made was fully bipedal, but this was clearly not the case.

In addition Lucy had shoulder sockets that faced upward, a common feature of modern apes. This unique feature enables apes to dexterously climb and swing from tree branches. In contrast, humans have downward facing shoulder sockets at birth that gradually develop to face forward as we become adults. This position is also integral to the uniquely human walking gait. Also in contrast to humans, ape shoulder morphology does not change during development. In other words living non human primates such as bonobos and australopithecines are probably similar in this regard, and therefore, neither can be considered any closer to humans than the other.

The important thing to appreciate is that all australopithecines including Lucy resembled apes more than they did humans. One can say that modern apes resembled humans today too but that does not make them human by any sense of the imagination. Today, the australopithecines are shown as an isolated evolutionary dead-end, an extinct branch in the tree of evolution as depicted in recent evolutionary family tree charts such as the one on display at the Smithsonian Museum. Interestingly, to cover all options the museum makes the comment that the "Species in the group [Australopicus Group] of early humans walked upright on a regular basis, but they still climbed trees too." Isn't that what bonobos do today?

Recently, this view has been confirmed in an article published in LiveScience under the heading, *Early Human 'Lucy' Swung from the Trees* that says, "The question as to whether Australopithecus afarensis was strictly bipedal or if they also climbed trees has been intensely debated for more than 30 years". According to the article, David Green at Midwestern University in Illinois, said that, "*These remarkable fossils provide strong evidence that these individuals were still climbing at this stage in human evolution.*" [7]

It is evident that Wheeler had begun his theory with the false premise based upon the evolutionary evidence at the time that our human ancestors (Lucy) was bipedal and lived in a savannah habitat 3.2 million years ago. To make matters worse, when DM Wilson and GD Ruxton of Glasgow University repeated Wheeler's experiment in 2011 updating them in line with new developments and measurements in animal thermal biology they came to a startling conclusion. By modifying the models to represent a running hominin rather than Wheeler's stationary forms the results ruled out Lucy altogether.

"Our model suggests that for endurance running to be possible, a hominin would need locomotive efficiency, sweating rates, and areas of hairless skin similar to modern humans. We argue that these restrictions suggest that endurance running may have been possible (from a thermoregulatory viewpoint) for Homo erectus, **but is unlikely for any earlier hominins**."* [8] [bold mine]

This means that Wheeler had started out on the wrong premise. He had assumed from what was written at the time that Lucy was said to have inhabited the equatorial savannah, and he based his theory on that basis. He was wrong! Unlike Lucy, Homo erectus was not confined to that kind of habitat so his theory was therefore invalid.

Homo erectus is described as an extinct species of hominin that lived throughout most of the Pleistocene, with the earliest first fossil evidence dating to around 1.8 million years ago and the most recent to around 200,000 years ago. [9] That is not a lot of time in evolutionary terms for major changes in the human anatomy to accommodate, bipedalism, hair loss and thermoregulation enhancements. And, according to the Natural History Museum website, Homo erectus was not confined to the hot savannah region that Lucy was suppose to have inhabited. "Migration happens for many reasons but essentially H. erectus probably drifted across northern Africa, across the Sinai region into Asia, when suitable habitats and food sources stretched that far. Meat was an important part of their diet and carnivorous animals often range more widely than herbivores. This, together with their larger body size, helps explain the large geographic range of H. erectus." [10]

MORE PROBLEMS

To recap, Wheeler's theory relied on a number of assumptions that he perceived was accurate. First he believed that Man's ancestors became bipedal about 3 million years ago, which would allow time for the major changes in the anatomy of Man to take place in order to accommodate becoming bipedal, resulting in his loss of hair and the development of thermoregulation to compensate for his nakedness. That time has now been cut down by half or even more making that transformation highly unlikely. His theory also assumed that Man had morphed into the naked creature that he had become because an erect posture was adopted by early hominids to combat high temperatures, especially when travelling exposed to direct sunlight as in the open savannah.

Wheeler's thermoregulatory hypothesis for bipedal origins, based on movement across patches of open grassland between clumps of forest, is simply not workable because these habitats would offer shade from the sun and shield the hominids from breezes - both assumed in his model. The question is why would hominids leave the shade in the hottest part of the day in equatorial zones? To postulate that hominids would have done this, as Wheeler suggests, in order to gain an advantage in foraging time against competitors, seems fanciful in the extreme and generates more questions than it answers.

For example, what benefit does an upright posture give a hominid, in terms of thermoregulation, when the sun is lower in the sky? And: If thermoregulation was achieved by sweat cooling, from where did these hominids replenish the water that would have been lost in this way? Assuming, as this model seems to, that early hominids always lived within close proximity to permanent fresh water supplies appears to defeat the premise on which the model is based: water side vegetation, even grasses, are usually much taller than 1.25 metres.

It could be argued that if hominids foraged around noon in equatorial habitats in waist high grasses, Wheeler's model could work. At such times potential food sources might be left alone by other predators and scavengers alike, preferring to keep cool in the shade. However, the selective benefit of having this two to three hour window of opportunity open to them would appear to be limited at best and, perhaps, be more than negated by an increased risk of overheating. However, today it is very rare for mammals to be observed foraging around noon, let alone any ape. Besides, much of the evidence emerging since Wheeler's papers indicate that early hominid environments were relatively wet and wooded and not arid and open.

On these criticisms alone it is clear that Wheeler's theory cannot stand up and while his experiments and calculations demonstrated the thermoregulatory advantages of bipedalism over quadrupedalism and the advantages of increased body size in hot savannah environments, the results do not indicate that the initial step in the denudation process occurred in open hot environments, or that bipedality preceded body hair reduction. Furthermore, his estimates of the percentage of total body surface area exposed to direct solar radiation merely demonstrate the advantage of bipedalism in relation to quadrupedalism.

Wheeler's theory ignores the fact that a covering of hair is a multifunctional mammalian asset; it would not become redundant merely because one function (heat regulation) was being otherwise catered for.

Furthermore, a reduced density of hairs per square centimetre does not necessarily entail decreased coverage. For example gorillas, the largest of the great apes, are larger than humans and their hairs are farther apart, but they are certainly not more naked. The mountain gorilla beringei has a particularly deep and luxurious fur coat. Thus, if a hominid had needed a better shield against the sun's rays on the hot savannah, body hairs would have grown longer rather than shorter.

One last thing that negates Wheeler's theory and it is the same one that Darwin found difficult to explain. In *Descent* Darwin wrote, "The fact, however, that the other members of the order of Primates, to which man belongs, although inhabiting various hot regions, are well clothed with hair, generally thickest on the upper surface on the head of man being covered with long hair; also on the upper surfaces of monkeys and of other mammals being more thickly clothed than the lower surfaces." [11]

Why would Man lose his hair when the other primates who inhabited the same region did not? Wheeler argues that this is because Man had become bipedal. The problem with this is that in environments where mammals are exposed to strong solar radiation, the coat acts as a shield, reflecting and reradiating heat before it reaches the skin. Hence, for most mammals, the loss of this insulation would create more problems that it would solve. Besides, we now know that Wheeler's premise that bipedalism took place because of the heat of the savannah is seriously flawed for the reasons already discussed and therefore the theory falls at the first hurdle. End of story!

CHAPTER 5
DESPERATE FANTASIES

"The evolution of near nakedness in the human species has been accounted for by a series of myths which owe more to the predilections of their creators than to the available evidence."
(Professor Francis Ebling,
"Journal of Human Evolution", January 1985)

THE PARASITE THEORY | THE CLOTHING HYPOTHESIS | THE NEOTENY HYPOTHESIS | THE ALLOMETRY HYPOTHESIS | THE FINAL ANALYSIS | THE CARRION-EATING HYPOTHESIS | THE HUNTING HYPOTHESIS

We humans as we have seen have been described as the naked ape, and our nakedness undoubtedly constitutes one of the most striking differences in appearance between humans and other apes. Although we are truly not hairless we are considered as such because the hairs that we do have are minuscule, and certainly neither protect the skin nor provide any appreciable thermal insulation. Therefore, explaining why we became naked has been a daunting challenge for evolutionists.

Thus far we have taken a look at the major theories that have been put forward, namely sexual selection, the aquatic ape theory and thermoregulation. All have failed miserably so this has led to the invention of some weird and fanciful theories to try to address the problem of our hairlessness. Under normal circumstances these theories would be laughable and would not be given a second look but the problem is that they have been suggested by a few eminent professors and their theories published in leading science journals. Consequently, some people actually believe what they say with almost religious fervour. Let us take a look at some of these theories.

THE PARASITE THEORY

In 1874 Thomas Belt (1832-1878), an English geologist and naturalist first introduced the idea that a naked primate would be less liable to harbour ticks and other noxious parasites, which, in the tropics, may constitute a serious danger to health. That may be so, but Belt does not explain how Man had become hairless because of this. Darwin in his book **Descent of Man** rejected the idea and quotes the words of Sir W. Denison to put Belt's hypothesis to bed.

> "*But whether this evil is of sufficient magnitude to have led to the denudation of his body through natural selection, may be doubted, since none of the many quadrupeds inhabiting the tropics have, as far as I know, acquired any specialised means of relief.*" [1]

In 1967 Desmond Morris in his famous book **The Naked Ape** raised the parasite subject again, and dismissed it for the very same reasons that Darwin did. Why did other primates retain their hair when Man's common ancestor did not?

The theory should have been buried for good, but then in 2003 professor Mark Pagel, at the University of Reading, and Sir Walter Bodmer, at the John Radcliffe Hospital in Oxford, resurrected the idea. They said in **Proceedings of the Royal Society** under the title "A naked ape would have fewer parasites" that, "*Our hypothesis explains features of human hairlessness - such as the marked sex difference in body hair, and its retention in the pubic regions - that are not explained by other theories.*"

What was that hypothesis? Humans, they say, lost their hair in order to reduce the burden of parasites such as fleas and ticks, some of which would have transmitted disease. They said that early humans probably lived close together in hunter-gatherer groups, in which the rate of parasite transmission was high. Hairless skin was easier to keep clean.

Cultural adaptations, such as the use of fire, shelter and clothing, allowed humans to become furless.

So let us get this right! As if by magic and waving a magic wand, because external parasites were a pain, evolution in its all knowing, all seeing randomness decided to divest humans of their hair. I thought evolutionists did not believe in miracles. Now listen to what Pagel next says as he expounds upon this fantasy. "*Once hairlessness had evolved through natural selection*", Dr. Pagel and Dr. Bodmer suggest, "*it then became subject to sexual selection, the development of features in one sex that appeal to the other. Among the newly furless humans, bare skin would have served, like the peacock's tail, as a signal of fitness.*" [2]

From what you have read in this book so far, doesn't what Pagel and Bodmer say sound ridiculous? It is no good calling upon natural selection to account for Man's hairlessness, as both Darwin and Wallace rejected this possibility. Nor can the two learned professors call upon sexual selection to help themselves out either for reasons explained elsewhere. In other words they are talking a lot of bull, if you pardon my expression.

Think about it. Just imagine for the sake of expediency that one of our hairy common ancestors was born with a bald patch on it's body. We will call him Joe. As Joe romped through the grasses of the savannah or swung through the trees, he would have been an easy target for parasite attack. Normally, parasites do not have free access to bare skin but have to burrow down through fur to gain access to their feeding grounds. So like chimpanzees today, our common ancestors spend a great deal of time grooming each other to rid themselves of these parasites. But poor Joe, his bare skin would be like waving a red flag to a bull. He would not have stood a chance, and before long his lovely bare bald patch would be a seething mass of parasitic invasion caked in dried skin and scarred with sores and boils.

Joe's chances of survival would be minimal and I doubt a female would be drooling over his "manliness". Why? It is interesting to note that many mite species, such as scabies mites and chiggers, produce acute skin irritations and for example, in livestock any persistent skin inflammation (often with accompanying hair loss) caused by mites can seriously weaken animals. Joe would hardly be the fit and rugged mate a female would fancy under such circumstances, assuming the sexual selection was a viable theory. Besides, grooming plays a key role in the social structure of a chimpanzee group and we can safely say that the same would have been for Man's so called common ancestor.

According to **Animal Planet** from the Discovery channel, adult chimpanzees spend about an hour each day in a friendly social activity called grooming. During this time, two or more chimpanzees sit and pick through each other's hair. They remove dirt, insects, leaves, and burs (the seeds of certain plants) from each other. In addition to helping keep each other clean, grooming reduces tension among group members. Chimpanzees are sociable animals who seem to need physical contact, and grooming helps satisfy this need. It also strengthens ties between group members.

Sometimes as many as ten chimpanzees will take part in these peaceful, relaxed grooming sessions. Often a grooming group will be made up of a mother and several of her offspring of different ages. Adult males may also groom one another. [3]

Under such circumstances, with each chimp being groomed by their fellows for lengthy periods and its importance as a friendly social activity that helps members of a group to bond with each other, why would our common ancestors need to become hairless? Would the random small incremental unguided processes of evolution kick in and cause our common ancestor to become hairless because it might have been easier to remove troublesome parasites from naked skin? Pull the other leg, it's got bells on it!

Pagel and Bodmer's parasite theory sounds all too familiar. It sounds like Lamark's famous giraffe analogy, long proven to be erroneous. He postulated that if animals began to use an organ more than they had in the past, it would increase in its lifetime. If a giraffe stretched its neck for leaves, for example, a "nervous fluid" would flow into its neck and make it longer. Its offspring would inherit the longer neck, and continued stretching would make it longer still over several generations. Meanwhile organs that organisms stopped using would shrink. The parasite theory uses the same faulty reasoning except that it is not the neck that is growing but the patch of bare skin that is expanding, because it was being scratched more often enabling parasites to be scraped away more easily.

One of the major arguments that Pagel and Bodmer used to support their hypothesis in their paper presented to the *Proceedings of the Royal Society* aforementioned was because they said that hairlessness was made possible in humans was because it owed, "...to their unique abilities to regulate their environment via fire, shelter and clothing. Clothes and shelters allow a more flexible response to the external environment than a permanent layer of fur and can be changed or cleaned if infested with parasites." From this it appears that the wearing of clothes was the key for Man's hairless condition. The clothing hypothesis was rubbished a long time ago.

THE CLOTHING HYPOTHESIS

The next suggestion for our hairlessness is probably the most farcical theory of them all. Apparently when we began to wear clothes, gradually over time we shed our hair. Why would our hairy ancestor want to wear clothes if it already had an insulative coat of body hair? Yet the suggestion is a serious one because it appeared in the prestigious magazine *Science* magazine in 1966. Under the heading of **Evolution of Hairlessness in Man**, Bentley Glass (1906 - 2005) an American geneticist and noted columnist wrote that hairlessness was the effect of clothing; and because a shirt opens in front, chest hair has remained. [4]

I am lost for words. Bentley Glass received his Ph.D. degree under the mentorship of geneticist Hermann Joseph Muller and held many distinguished academic titles, so he is person worthy of note. But to come up with this lame theory to explain human hairlessness is beyond belief.

From an evolutionist viewpoint, recent DNA analysis by Rogers, Iltis & Wooding in *Current Anthropology*, suggests that humans have been hairless savannah dwellers for at least 1.2 million years. Concluding their report and also referencing the parasite theory of Pagel they wrote:

> *"Finally, some argue that hairlessness evolved in response to clothing (Glass 1966, Kushlan 1985). The most recent variant of this argument (Pagel and Bodmer 2003) holds that hairlessness evolved to reduce parasite load, an adaptation that was made*

feasible by clothing and control of fire. There is no evidence of tailored clothing before about 20,000 years ago (Klein 1999:536) or even of hide scraping before 300,000 years ago (Toth and Schick 1993:161). **Thus, the present results indicate that humans were naked before they were clothed."** [5] [bold mine]

If that was not enough from a historical perspective cuneiform records describe a time when Man was naked before becoming involved in agriculture, 4500 years ago. As the *The Oxford Handbook of Cuneiform Culture* notes:

"The earlier texts cited above are hardly less explicit, but there civilisation [Sumerian] is thought to have come directly from the gods without the sages' intervention: **beforehand people were naked** *and ate grass like sheep; afterwards were dressed (Ewe, flax), produced grain (Wheat), and ate bread."* [6] [bold mine]

Furthermore, we see numerous examples of the daily life of the ancient Egyptians depicted in ancient frescoes in Sakarra dated about 2340 BC showing naked men going about their business - fishing and farming.

All the evidence points to the fact that humans were naked long before they started to wear clothes. **People wore clothes because they were naked, not because wearing them made them naked.** Bentley Glass could not have been more wrong. Once again another desperate attempt to explain Man's hairlessness falls by the wayside.

The next theory that we shall now discuss is one that has been proposed by Stephen Jay Gould (1941-2002), an American palaeontologist, evolutionary biologist, and historian of science. It is called the Neoteny hypothesis and perhaps, because of Gould's backing, it is one theory that should be looked at seriously.

THE NEOTENY HYPOTHESIS

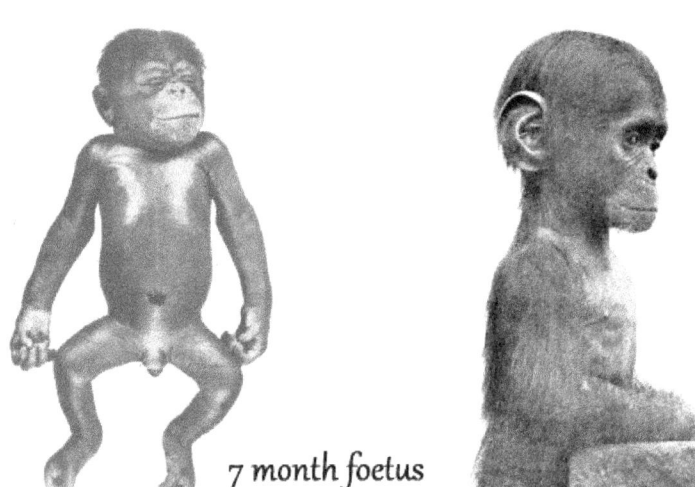

7 month foetus   Infant

Derived from the Greek word "neteinein" means "holding onto youth" the term 'neoteny' describes the retention of juvenile features of an ancestor in adult forms of the descendant. Neoteny is also called juvenilisation and it has been suggested that humans are a juvenilised form of ape. The chief proponent of the theory is that of Stephen Jay Gould who was one of the most influential and widely read writers of popular science of his generation. Gould spent most of his career teaching at Harvard University and working at the American Museum of Natural History in New York. In the later years of his life, Gould also taught biology and evolution at New York University.

In 1977 Gould published a definitive account of neoteny in his book *Ontogeny and Phylogeny* and it was hailed a classic in his day. Even so, it is significant that the matter of hair loss was barely touched upon in the book.

When describing neoteny, Gould found it convenient to consider the chimpanzee as representative of a human-like forebear perhaps five million years old. A chimpanzee foetus at seven months shows body hair on the scalp, eyebrows, borders of the eyelids, lips and chin, precisely those places where hair is retained in adult humans. Then once born a chimpanzee infant has a relatively small chin, large head relative to body size, large eyes relative to face and upright positioning of the head on the neck. In fact, according to Gould, it would appear that a chimpanzee infant exhibits a number of features much like that of a contemporary human adult.

Gould said that in order to accommodate our brainy super-sociality, we needed to extend our brain developmental period for nearly two decades. After which our brains constantly break down and only form new connections from existing neurons as we learn. Apparently, one of the easiest ways to this state was to push back our total maturation, making our adults more and more like large versions of a chimpanzee foetus. In fact, our development with the womb takes so long that a large chunk of it needs to be done outside the womb. Hence, all human babies are born premature (although some more than others).

In plain English what Gould and others have said is that modern humans were not humans at all but rather apes stuck in a juvenile state of development, a primate foetus that has become sexually mature. Gould therefore calls humans "neotenic apes". He said that humans were characterised by a general slowing down (retardation) of pace of development, so that they mature more slowly than other primates and therefore live longer. Hence, humans retained the relatively large head of the infant chimpanzee and also the hairlessness of the foetal ape. If the hairlessness of the foetal ape was being retained into adulthood by a process of neoteny, one would expect the human body to retain this characteristic throughout its whole development from embryo to adult. Hence, that is why Man is almost hairless.

To say that we are neotenic apes is incredulous to say the least but is there any justification to support such an outrageous claim? Gould used a seven month chimpanzee foetus as evidence of his claim because it shows body hair on the scalp, eyebrows, borders of the eyelids, lips and chin that are found in adult humans. However, if the hairlessness of the foetal ape was being retained into adulthood by a process of neoteny, one would expect the human body to retain this characteristic throughout its whole development from embryo to adult. However, this is not the case. A six month human foetus is completely covered with a coat of very fine, soft, and usually unpigmented, downy hair known as lanugo. It is normally shed before birth, around 7 or 8 months of gestation but is sometimes present at birth and disappears on its own within a few days or weeks.

Lanugo hair will invariably be shed by three to four months after birth. It is replaced by hair covering the same surfaces called vellus hair, but this hair is finer and more difficult to see. The more visible hair that continues into adulthood is called terminal hair. At puberty, the androgen hormone causes much of the vellus hair to turn into terminal hair and stimulates the growth of new hair in the armpit and the pubic area. In men, this change in vellus hair occurs on the face and the body. What all this means is that the human infant is already hairless at birth and does not become hairless as it grows, while the opposite is true of the chimpanzee infant. Hence, the juvenile characteristics that Gould speaks being retained as a human grows simply do not apply. [7]

Professor Markus Rantala of the Department of Biology at the University of Turku, Finland describes another weakness to the theory.

> *"Another weakness of the theory [neoteny] is that while some characteristics may be retained as part of a neotenic package, this only applies to characteristics that are either benign or neutral in their effect on fitness to survive. No one claims that all foetal characteristics are retained in a neotenic species. For example, a human foetus and a human baby both have very short bandy legs, but natural selection ensures that this feature is not retained in adult life (Morgan, 1990). Furthermore, the neoteny theory does not tell us anything about the value of nudity as a new character that helped the naked ape to survive better in his hostile environment (Morris, 1967)."* [8]

Gould claims that because of the growth requirements for our large brains we needed to extend our brain developmental period for nearly two decades and that is why we evolved the retention of juvenile features. This is nonsense of course as Professors Robson of Department of Anthropology, University of Utah, Salt Lake City and Bernard Wood of the Centre for the Advanced Study of Hominid Paleobiology, Department of Anthropology, The George Washington University in Washington DC explain. Writing in the *Journal of Anatomy* in 2008 they said that,

> *"Similarities between chimpanzees and modern humans do not support the view that our juvenility is longer because of the growth requirements of our large brains. Whereas adult brain size is strongly correlated with the length of subadulthood (Leigh, 2004), age at brain growth cessation is not. These data show that encephalization in primates is achieved through an increased velocity, not longer relative duration, of brain growth and challenge the widely held assumption that the length of brain growth is linked to, and sets the pace of, life history."* [9]

Dr Andrew Lock (Massy University, New Zealand) and Dr Charles Peters (University of Georgia, Georgia. USA) also said that it was a misconception that human development is generally retarded or slow across the life span. *"Humans do not grow more slowly than other primates but grow for a longer time in each phase of growth"*. [10] And Dr Kenneth McNamara, Director of the Sedgewick Museum of Earth Sciences, University of Cambridge in his book *Shapes of Time* explains in detail how the relative expansion in each phase of human development has been misunderstood as retardation or neoteny.

> *"I believe there has been a basic mistake in equating delays in transition from one growth rate to another with reduction in growth rate. They are nothing of the sort.* ***Had humans been the product of reduced, neotenic growth we would be vastly different beasts, small of stature, small of limb, and, significantly, small of brain.*** *Delays in transition from one growth phase to the next and neoteny are emphatically different processes, yielding fundamentally different results. In the case of humans the product of our pattern of development, which are characterised by long, drawn out growth phases, is overwhelmingly not one of paedomorphosis but one dominated by permorphosis. In many important ways we have developed 'beyond' our ancestors and all other primates."* [11] [bold mine]

For the reader's information permorphosis is where individuals of a species mature past adulthood and take on hitherto unseen traits, while paedomorphosis is the retention by an organism of juvenile or even larval traits into later life. Once the reader has got past the scientific mumbo jumbo it is evident that, although Gould was an eminent scientist, just because he declares to support a theory, does not make it so. We are humans and not neometic apes and our near hairless condition cannot be justified by calling upon a theory that leaves much to be desired and where the mechanisms concerning it are widely disputed.

THE ALLOMETRY HYPOTHESIS

Here we have another unfamiliar term 'allometry' to contend with. What the term refers is the fact that as species become larger not all organs of their bodies increase in the same ratio as their overall body size and mass. This led to professors Gary Schwartz and Leonard Rosenblum in 1981 to publish a theory to account for Man's hairlessness in the *American Journal of Physical Anthropology* under the heading **Allometry of primate hair density and the evolution of human hairlessness**.

The two scientists reasoned that a primate that is twice the size of another primate does not have twice as many hairs to cover the increased surface of its body. As supportive evidence they carried out allometric analyses of hair densities in twenty-three anthropoid primate taxa to reveal that increasingly massive primates have systematically fewer hairs per equal unit of body surface. As an example, they pointed to a small primate like the marmoset whose hairs grow very close together, but the larger the primate, the fewer hairs it has per unit of body surface. [12]

This sounds quite reasonable at first glance and on this basis therefore the allometric theory proposes that as the first hominids descended from apes and since apes are larger than monkeys, the hominids would have had sparser hair. That would mean that they would be less well protected against the direct heat of the sun than small savannah species like baboons and patas monkeys. It is therefore further suggested that hominids had to evolve an alternative system of keeping cool, that is, perspiration and this system ultimately became so effective that they were able to dispense with body hair altogether. As if dispensing with body hair was something the hominids had any control over. This is another example of evolutionists waving their magic wand to make things happen to suit their theories.

There are big problems with this theory. A reduced density of hairs per square centimetre does not necessarily entail decreased coverage. If that was the case how can one explain the hair covering of gorillas. Gorillas are the largest of the great apes and are even larger than humans and while it is true that their hairs are farther apart, they are certainly not more naked. The mountain Gorilla beringei has a particularly deep and luxurious fur coat. Thus, if a hominid had needed a better shield against the sun's rays on the hot savannah, its body hairs would have just as likely grown longer rather than shorter, just as gorillas show. Besides, the allometric theory ignores the fact that a covering of hair is a multifunctional mammalian asset; it would not become redundant merely because one function (heat regulation) was being otherwise catered for.

Finally, to put this ridiculous theory where it belongs, to the funny farm, human functional nakedness is not caused by hair follicles being more widely spaced. It is caused by having follicles that produce hair so short and fine that they are not even visible to the naked eye: some are too short to reach the surface of the skin. There is no evidence of a trend in living primates towards shortening of body hair, whether in response to body size or any other variant.

THE FINAL ANALYSIS

The aforementioned theories do not offer any viable solution to why humans have naked skin, and although they are sometimes mentioned in evolutionist literature, nobody really takes them seriously. Upon reading this chapter you can understand why.

Other theories are even more bizarre and I only mention a few here in passing, just to show how desperate evolutionists have become in trying to answer why, as Man evolved he lost his hair. **The Carrion-Eating Hypothesis** really takes the biscuit. The theory is based on the observation that the messiest eaters in the animal kingdom are the vultures and condors, which feed on carrion, and guess what? They have naked necks!

So it was that William Stephensson proposed the theory in a book he wrote in 1972 called *The ecological development of man*. The theory was based upon the treatment of Aboriginal culture towards food, food collection, hunting and Aborigine eating habits. It showed that in this particular case humans were messy eaters and so it would be helpful if they were naked, just as Aborigines in the past had been. [13] Really? As the famous tennis player John McEnroe used to say, "you cannot be serious?"

Then there is **The Hunting Hypothesis**. In fact for a while this theory was a serious contender to the present popular thermoregulation theory as discussed in the previous chapter. It was proposed in 1966 by Bernard Campbell who was professor of anthropology at the University of California, Los Angeles. His book *Human Evolution: An Introduction to Man's Adaptations* tried to answer why one savannah primate needed to go naked while all other species in the same habitat retained their fur. This was explained on the grounds that vegetarian primates do not need to move very fast; a carnivorous primate, on the other hand, would get hot while chasing its prey, and losing its hair would enable it to cool down. [14]

Desmond Morris also considered such hairlessness necessary because hunting chases were activities for which early man would be otherwise physically poorly adapted. However, both Campbell and Morris failed to grasp the major flaw in the theory. It was the male who was supposed to be the hunter and allegedly became overheated in the chase across the hot savannah, but it was the female who according to Darwin and others who became the most hairless by remaining in camp bringing up the children. Besides, one only has to look at the present animals hunters hunting in the savannah today such as the cheetah and observe that none are naked.

While evolutionists continue to struggle to get their heads around explaining how humans had evolved to become the naked almost hairless creatures that they are, news in the field of genetics has bolstered their belief in a common ancestor between chimpanzees and Man. This has diverted attention from their insoluble problem about our hairlessness, because it was declared that chimpanzees share nearly 99 percent of our DNA. That being the case, it is reasoned, then just because a tangible explanation for the Man's naked skin has not been found yet, that does not mean that there is no explanation because our DNA is so close to chimpanzees that it must have happened. But is it really true that humans share nearly 99 percent of the DNA with chimpanzees or are we looking at a red-herring? This is the topic of the next chapter.

CHAPTER 6
THE DNA GAME

"Chimpanzees are the closest living relatives of humans and share nearly 99 percent of our DNA."
(Scientific American, 20 April 2009)

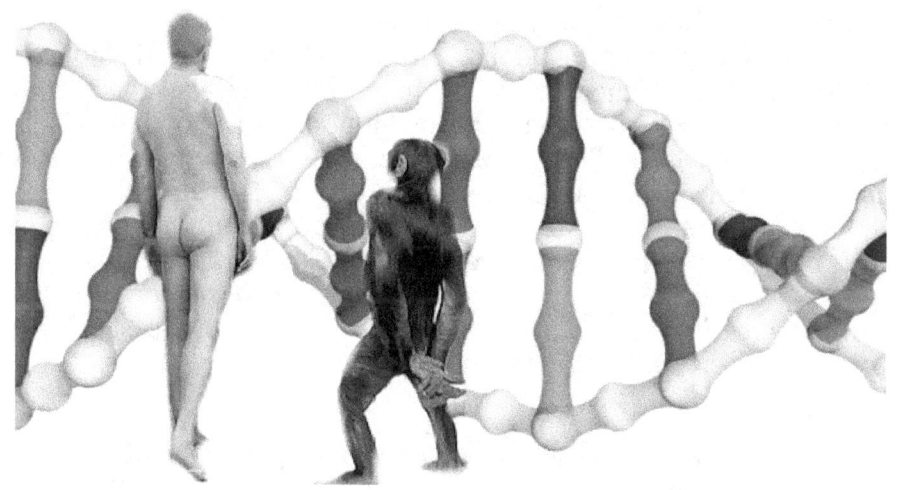

WHAT IS DNA? | THE HUMAN GEMONE PROJECT | THE MAKING OF A MYTH | IF TRUTH BE TOLD | THE X AND Y CHROMOSONE PARADOX | CHROMOSOME 2 FUSION THEORY | WHAT DOES ALL THIS MEAN?

It seems that everyone knows that 98% or 99% of the DNA in chimpanzees were found to be identical in humans. This is because evolutionists have done a fantastic job with their propaganda and leading the field is none other that Richard Dawkins. As far back as 1986 in his book *The Blind Watchmaker* he stated that *"Chimpanzees and we share more than 99 per cent of our genes."* [1]

Once Dawkins had expressed this "fact" and made it public, the incredible similarity between chimpanzee and human DNA became one of evolutionists favourite weapons to whack on the head of anybody who disagreed with evolution, and according to them this was decisive evidence of the common ancestry of apes and humans. However, what is not commonly known is that Dawkins was really expressing a view that was based on a very small amount of data, **a partial map of DNA** that had been mapped out in 1974.

The truth of the matter is this. The genome map (the entire map of an organism's hereditary information) for human and chimpanzees had not yet been done and would not be completed until almost twenty years later. Initial drafts of the human genome were finished in 2001 but it would not be for another four years, in 2005 when the chimpanzee genome was completed as well. Dawkins had in fact misled the entire scientific community with his statement. One cannot make a "fact" of science when it is based on only partial data.

Even when the entire human genome had been mapped and before the chimpanzee genome had been done there was doubts about the 99% claim that Dawkins and others were saying. For example, the *National Geographic Magazine* in 2002 published a news article by David Nelson, a geneticist at Baylor College of Medicine in Houston under the heading **Humans, Chimps Not as Closely Related as Thought?**. The article said, "*For decades, scientists have agreed that human and chimpanzee DNA is 98.5 percent identical. A recent study suggests that number may need to be revised. Using a new, more sophisticated method to measure the similarities between human and chimp DNA, the two species may share only 95 percent genetic material.*" [2]

So we have a new figure of 95% similarity between chimpanzee and human DNA, which is embarrassing for evolutionists because according to researchers in Poland from the Institute of Zootechnics in Kraków, "*Pigs are very similar to humans in terms of their DNA makeup. Some 94 percent of pig DNA matches with the DNA of a human being.*" [3] That is why these researchers are working with genetics on pigs to create human body part for transplants.

With only 1% difference between the DNA of a pig and a chimpanzee it is funny that evolutionists are not rushing around promoting the idea that Man may have evolved from pigs. However, one of their number has not been deterred. His name is Dr. Eugene McCarthy and he is a leading geneticist who has made a career out of studying hybridisation in animals. He has amassed an impressive body of evidence suggesting that human origins can be best explained by hybridisation between pigs and chimpanzees. [4]

McCarthy says that, "*when he searched the literature for traits that distinguish humans and chimps, and compiled a lengthy list of such traits, he found that it was always humans who were similar to pigs with respect to these traits.*" [5]

Needless to say he has come under attack from all directions, creationists and evolutionists alike, but most criticisms have been directed towards McCarthy himself rather than his theory. *"By and large, those coming out against the theory had surprisingly little science to offer in their sometimes personal attacks against McCarthy."* [6]

Are you confused? You should be. What really is the truth of the matter? Could it simply be a case as Mark Twain once said, "There are three kinds of lies: Lies; damn lies; and statistics." Scientists can present whatever numbers they want to emphasise whatever they want. Which numbers are the most important is really a matter of opinion depending upon who is making the assessment.

For example, it is in evolutionists best interests to always present the numbers in such a way as to diminish the difference between humans and chimpanzees in order to make it more plausible that there is a common ancestor. And as we shall see, this is exactly what Dawkins and others have done, and once again we find the public fooled by unsubstantiated evolutionist rhetoric.

To get to the bottom of this often heated topic of conversation I think it would be appropriate to describe what DNA is so the reader can understand some of the terms used by geneticists mentioned in this chapter and make an informed judgement.

WHAT IS DNA?

Every living thing is made up of cells and humans are no exception. In fact it is estimated that we have trillions of cells that make up our body, with each cell containing 46 (23 pairs) chromosomes packed tightly into the region of a cell called a nucleus. It is interesting to note that apes such as chimpanzees, bonobos, gibbons and gorillas have 48 (24 pairs) chromosomes, which is two more than humans. This in itself should give you cause for thought to the question whether or not Man evolved from a common ancestor. We shall look into this in more detail later, but for the time being, back to the plot.

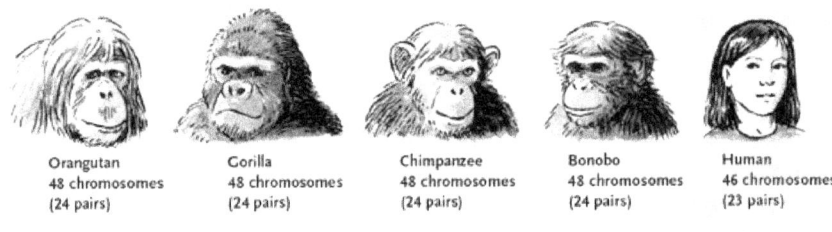

Orangutan	Gorilla	Chimpanzee	Bonobo	Human
48 chromosomes	48 chromosomes	48 chromosomes	48 chromosomes	46 chromosomes
(24 pairs)	(24 pairs)	(24 pairs)	(24 pairs)	(23 pairs)

Apes such as chimpanzees, bonobos, gibbons and gorillas have 48 (24 pairs) chromosomes. Humans have 46 (23 pairs)

Chromosomes are the basic building blocks of life where the entire genome of an organism is essentially organised and stored in the form of DNA (deoxyribonucleic acid) which is present inside every cell making up that organism. A chromosome is made up of a single chain of DNA that forms a double helix. Each chromosome has what is known as a **telomere** at each end, and it protects the end of the chromosome from deterioration or from fusion with neighbouring chromosomes.

Many age-related diseases are linked to shortened telomeres and some scientists have suggested that telomeres determine how long we live. This is based upon the theory that a telomere is about 15,000 bases long at the moment of conception in the womb, but immediately after conception as our cells begin to divide, the telomeres begin to shorten each time the cell replicates itself. Once our telomeres have been reduced to about 5,000 bases, it is believed that this is when we die of old age. Consequently, as the theory goes, if our telomeres remained the same size regardless of how many times our cells replicated themselves, we would live forever. Who knows? Maybe the fruit of the Biblical tree of life contained a substance that prevented the resizing of the telomeres.

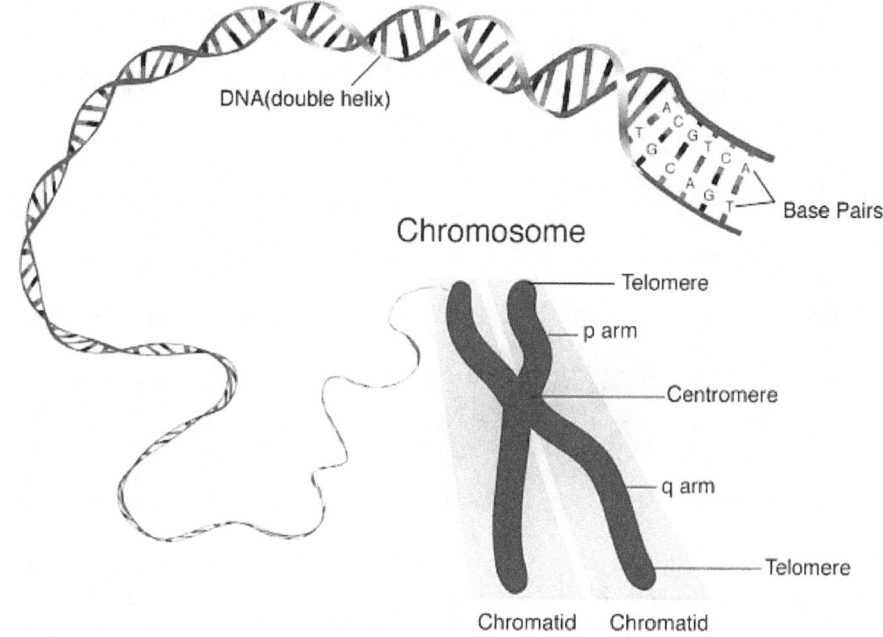

DNA(double helix)

Base Pairs

Chromosome

Telomere

p arm

Centromere

q arm

Telomere

Chromatid   Chromatid

DNA is in effect a genetic instruction book for enabling the production of proteins and cell processes that are essential to life and inherited from generation to generation. Every piece of DNA is composed of gene sequences containing instructions for each cell's development, reproduction and ultimately death. These instructions in the genetic instruction book takes the form of biological coding called **nucleotides**, which are very much like the digital coding of computers. Computer coding is based on a 2 digit code (zero and one), but the instructions in DNA uses a 4 digit chemical code called **bases.** These four bases comprise chemicals cytosine (C), thymine (T), adenine (A), and guanine (G). One of these bases connected to a sugar phosphate is called a **nucleoside.**

Nucleotides are arranged in two long strands that form a spiral called a double helix. The structure of the double helix is somewhat like a ladder, with the base pairs forming the ladder's rungs and the sugar and phosphate molecules forming the vertical side pieces of the ladder. The helix is further organised into short segments of DNA called genes. If you imagine DNA being a cookbook, then genes are the recipes within that book. Written in the DNA alphabet - A, T, C, and G - the recipes tell cells how to function and what traits to express. For example, if you have curly hair, it is because the genes you inherited from your parents are instructing your hair follicle cells to make curly strands.

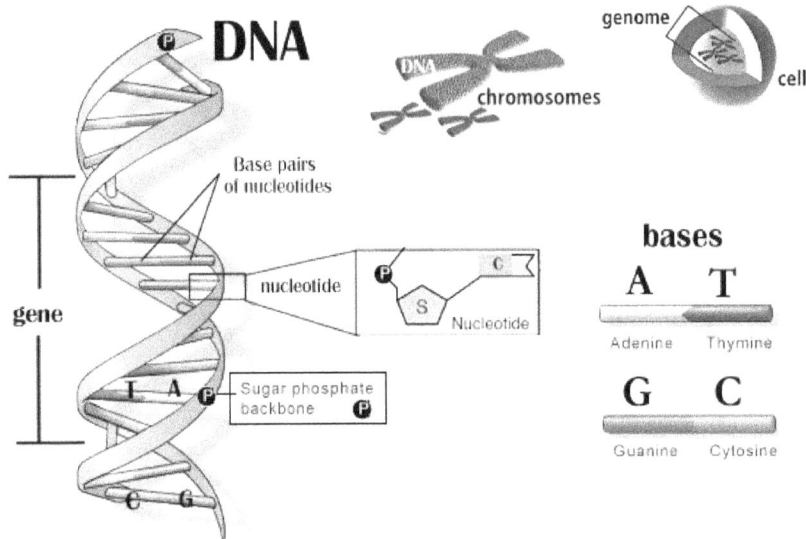

Cells use the genetic recipes written in our genes to make proteins - just like we use recipes from a cookbook to make lunch. Proteins do much of the work in our cells and your body as a whole. Some proteins give cells their shape and structure. Others help cells carry out biological processes like digesting food or carrying oxygen in the blood. Using different combinations of the DNA alphabet - As, Cs, Ts and Gs - DNA creates the different proteins - just as we use different combinations of the same ingredients to make different meals.

Cells come in a dizzying array of types; there are brain cells and blood cells, skin cells and liver cells and bone cells. But every cell contains the same instructions in the form of DNA. So how do cells know whether to make an eye or a foot? The answer lies in intricate systems of genetic switches. Master genes turn other genes on and off, making sure that the right proteins are made at the right time in the right cells. In order for DNA to create the different proteins it uses the nucleic acid present in all living cells called RNA (Ribonucleic acid) to act as a **messenger** for carrying instructions from DNA to control the synthesis of proteins.

RNA has the bases adenine (A), cytosine (C), guanine (G), and uracil (U) and RNA and comes in a variety of different shapes. The main job of RNA is to transfer the genetic code need for the creation of proteins from the nucleus to the ribosome. This involves transcription, decoding, and translation of the genetic code to produce proteins.

The ribosome is a complex molecular machine likened to a microscopic factory found within all living cells, that serves as the primary site of biological protein synthesis called translation.

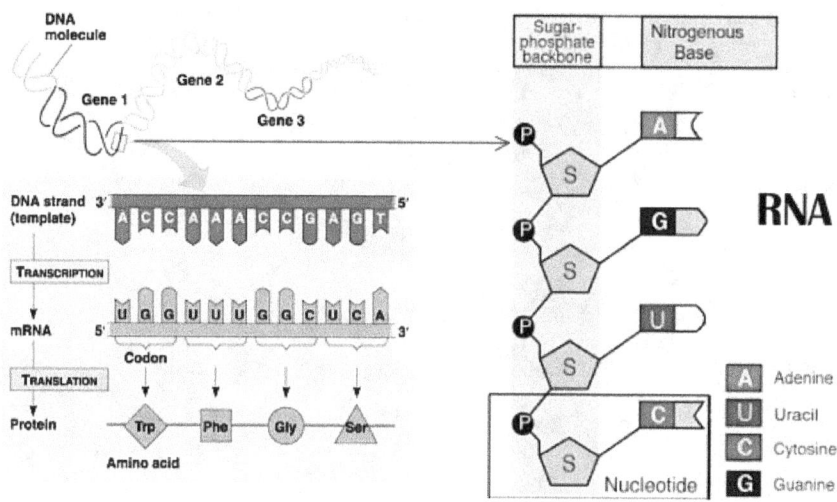

Ribosomes are found in many places around the cell and can be found floating in the cytoplasm, the clear gel-like substance enclosed within the cell membrane. These floating ribosomes make proteins that will be used inside of the cell. Other ribosomes are found attached on the Endoplasmic reticulum, a network of tubules and flattened sacs that serve a variety of functions within the cell. Those attached ribosomes make proteins that will be used inside the cell and proteins made for export out of the cell.

A ribosome consists of ribosomal proteins and rRNA (Ribosomal RNA) and are typically composed of two subunits: a large subunit and a small subunit. Scientists named them 60-S (large) and 40-S (small). When the cell needs to make protein, a message mRNA is created in the nucleus. The mRNA is then sent into the cell and the ribosomes. When it is time to make the protein, the two subunits come together and combine with the mRNA. The subunits lock onto the mRNA and start the protein synthesis. Another nucleic acid lives in the cell - tRNA, which stands for transfer RNA. tRNA is bonded to the amino acids floating around the cell. With the mRNA offering instructions, the ribosome connects to a tRNA and pulls off one amino acid. Slowly the ribosome makes a long amino acid chain that will be part of a larger protein, which will eventually be created in its final form.

This then is a brief guide of DNA. With this information we can now move forward and take an informed look at claims made by evolutionists that chimpanzees are the closest living relatives of humans, share nearly 99 percent of our DNA and which supposedly confirm the evolutionary hypothesis that Man and apes descended from a common ancestor. However, first we need to learn a bit about how the genome, the entire DNA structures of humans and chimpanzees, have been mapped.

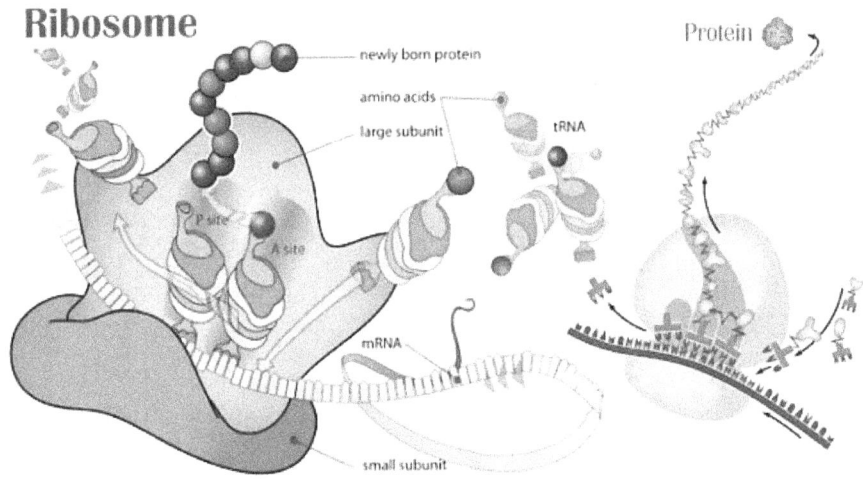

THE HUMAN GEMONE PROJECT (HGP)

The Human Genome Project (HGP) was an international, collaborative research program whose goal was the complete mapping and understanding of all the genes of human beings. As you have learned above, in the nucleus of nearly every human cell is a set of 46 chromosomes. These chromosomes are numbered in pairs from 1 to 22 for a total of 44. Each of the two remaining chromosomes are either X or Y. The numbered chromosomes are referred to as autosomes, and the X and Y chromosomes are called sex chromosomes. The human genome contains over 3 billion base pairs and these are made up of stretches of the four bases - cytosine (C), thymine (T), adenine (A), and guanine (G) - arranged in different ways and in different lengths. The average length of a gene is 27 thousand base pairs, but the largest known gene has 2.4 million base pairs.

All our genes in the DNA are known collectively as the genome. **The International Human Genome Sequencing Consortium** published the first draft of the human genome in the journal **Nature** in February 2001. However, at that time only 90% of the genome had been mapped. The full sequence was only finally completed and published in April 2003. [7]

There are trillions of cells in the human body and each perform a specific function depending upon what some gene sequences within the DNA has been programmed to do. Hang on a moment! Are we to believe that **The International Human Genome Sequencing Consortium** mapped all the DNA in all the trillions of cells that make up the human body? No! That would be impossible. Only the DNA of a single cell is mapped! Surely, you may ask, one cell in a trillion is hardly a great sampler to base an entire research project on. However, as Roseanne Zhao of the HGS says, "As the fundamental unit of life, each cell contains a complete copy of an organism's genome, which can undergo dynamic DNA mutations as the cell grows and divides." [8]

In other words, although only a single cell has been mapped, it does contain the same DNA sequences as all cells in the same organism. Consequently, if we are comparing the DNA of one human cell with another human cell they will never match one hundred percent. **No other human being on the planet share the exact same DNA formation**. We get part of our DNA makeup from our parents. However, even though our parents pass on their genes to their offspring they do not pass the exact formation on to every child. Despite being closely related, we still have our own special formation of DNA. That is why siblings in the same family can look very similar or very different depending upon the pattern of genes passed on to them at the time of conception.

By sampling the DNA from one cell belonging to one human this does not mean that it will be the same as that of another human. There will be some differences. Which leads to the problem concerning the DNA of chimpanzees and humans. If it is claimed that the DNA of a chimpanzee genome is a 99% match to that of a human genome, this suggests that for all intents and purposes they are a perfect match, after all what is 1% to be concerned about? This is impossible of course because humans and chimpanzees are physically and mentally different from each another, so that the claim of 99% cannot possibly be right. Clearly, something is seriously wrong with this figure when we look at this from a purely logical point of view. We therefore need to take a closer look at the chimpanzee genome to see if what is being compared is like for like.

After the completion of the Human genome project in 2003, the common chimpanzee genome project was initiated. Draft details of the chimpanzee genome was published in **Nature** on September 1, 2005, under the title "Initial sequence of the chimpanzee genome and comparison with the human genome." The report was produced by a group of genome scientists of the **Chimpanzee Sequencing and Analysis Consortium**.

It should be noted that the report assumes that human and chimpanzee species diverged from a common ancestor, and therefore the authors of the report did not look at their findings from a purely objective viewpoint. Consequently, one has to take what the authors say with considerable caution in view of the their evolutionist prejudices.

The summary of the report is as follows:

> "*Here we present a draft genome sequence of the common chimpanzee (Pan troglodytes). Through comparison with the human genome, we have generated a largely complete catalogue of the genetic differences that have accumulated since the human and chimpanzee species diverged from our common ancestor, constituting approximately thirty-five million single-nucleotide changes, five million insertion/deletion events, and various chromosomal rearrangements. We use this catalogue to explore the magnitude and regional variation of mutational forces shaping these two genomes, and the strength of positive and negative selection acting on their genes. In particular, we find that the patterns of evolution in human and chimpanzee protein-coding genes are highly correlated and dominated by the fixation of neutral and slightly deleterious alleles. We also use the chimpanzee genome as an outgroup to investigate human population genetics and identify signatures of selective sweeps in recent human evolution.*" [9]

Despite, their prejudicial evolutionary bias, the authors of the report are honest enough to say that "*The draft chimpanzee sequence here is sufficient for initial analyses, but* **it is still imperfect and incomplete**." [10] [bold mine] This is because only 94% of the chimpanzee genome was mapped [11] and as a consequence no percentage comparisons were made between chimpanzee and human DNA in the report. So it is very surprising to find that when newspapers such as the *Washington Post* published a news item about the study they wrongly stated that "As predicted by preliminary studies, the human and chimpanzee genetic codes are essentially 99 percent identical, a testament to how fundamentally similar the two species remain."[12]

What is going on? The report says nothing of the kind. If you doubt what I have said you are at liberty to read the full report at the link below.

www.genome.gov/Pages/Research/DIR/Chimp_Analysis.pdf

You will see that nowhere in the document do the authors claim that human and chimpanzee genetic codes are 99% identical. So where did the newspaper get its information? The clue is in the picture that accompanied the article.

The picture shows a chimpanzee and beneath it is written, "*The chimpanzee Clint's genome sequence was determined by scientists. Comparatively, it is almost 99 percent identical to that of humans. (Yerkes National Primate Research Centre)*" When we track down the research centre references it takes us to the *The International Human Genome Sequencing Consortium* website. This website published a review article entitled, **New Genome Comparison Finds Chimps, Humans Very Similar at the DNA Level** that was published by *NIH News* (US Department of Health and Human Services) which discusses the aforementioned report. It is from this article that we read:

The Washington Post

Scientists Complete Genetic Map of the Chimpanzee

By Rick Weiss
Washington Post Staff Writer
Thursday, September 1, 2005

Scientists said yesterday that they have determined the precise order of the 3 billion bits of genetic code that carry the instructions for making a chimpanzee, humankind's closest cousin.

The fresh unraveling of chimpanzee DNA allows an unprecedented gene-to-gene comparison with the human genome, mapped in 2001, and makes plain the

The chimpanzee Clint's genome sequence was determined by scientists. Comparatively, it is almost 99 percent identical to that of humans. (Yerkes National Primate Research Center)

"*The consortium found that the chimp and human genomes are very similar and encode very similar proteins. **The DNA sequence that can be directly compared between the two genomes is almost 99 percent identical.** When DNA insertions and deletions are taken into account, humans and chimps still share 96 percent of their sequence. At the protein level, 29 percent of genes code for the same amino sequences in chimps and humans.*" [13] [bold mine]

What we see is a kind of chain letter scenario, where the *Washington Post*, not having read the *Chimpanzee Sequencing and Analysis Consortium* report, instead reads the summary review of the report in *NIH News*, without realising that the report makes no mention that 99% of chimpanzee and human DNA is identical.

Furthermore, what we read in the review article is a very confusing picture. We are told that 99% of chimpanzee and human DNA is identical, but then when DNA insertions and deletions are taken into account that figure drops to 96%. And then, at the protein level, there is only 29% of gene code for the same amino acid sequences.

What is even more confusing is that having made such statements *NIH News* admits that, **"Despite the many similarities found between human and chimp genomes, the researchers emphasized that important differences exist between the two species**." [13] [bold mine] What are those important differences of which the editors of *NIH News* speak?

> "*About 35 million DNA base pairs differ between the shared portions of the two genomes, each of which, like most mammalian genomes, contains about 3 billion base pairs. In addition, there are another 5 million sites that differ because of an insertion or deletion in one of the lineages, along with a much smaller number of chromosomal rearrangements. Most of these differences lie in what is believed to be DNA of little or no function. However, as many as 3 million of the differences may lie in crucial protein-coding genes or other functional areas of the genome.*"

The DNA described as "*of little or no function*" is called Noncoding DNA (also called junk DNA) and it is defined as all the DNA sequences within a genome that are not found within protein-coding exons, and so are never represented within the amino acid sequence of expressed proteins. By this definition, more than 98% of the human genomes is composed of ncDNA. In other words only 2% of the genome is involved in producing proteins, so when *NIH News* says that there is at the protein level "only 29% of gene code for the same amino acid sequences", it is this 2% of the genome that is being referenced. Hence, when protein producing bases of human DNA is compared with that of chimpanzees, only 29% are the same. Are you getting this?

What then is the meaning of the 98% of the human genome that is composed of ncDNA? Recent books by Richard Dawkins, Francis Collins and others have used this "junk DNA" as evidence for Darwinian evolution and evidence against intelligent design (since an intelligent designer would presumably not have filled our genome with so much garbage). But recent genome evidence shows that much of our non-protein-coding DNA performs essential biological functions so that what Dawkins and the other writers have said in their books is rubbish, or should I say junk science.

For example, in 2009 Dawkins wrote, "*It stretches even their creative ingenuity to make a convincing reason why an intelligent designer should have created a pseudogene - a gene that does absolutely nothing and gives every appearance of being a superannuated version of a gene that used to do something -- unless he was deliberately setting out to fool us...Leaving pseudogenes aside, it is a remarkable fact that the greater part (95 percent in the case of humans) of the genome might as well not be there, for all the difference it makes.*" [14]

Now we know that Dawkins was wrong, very wrong. The intelligent designer did not make 95% of the human genome to do nothing. "Long stretches of DNA previously dismissed as "junk" are in fact crucial to the way our genome works" says the Guardian newspaper in 2012. Under the heading, "Breakthrough study overturns theory of 'junk DNA' in genome", **The international Encode project** said:

> "*For years, the vast stretches of DNA between our 20,000 or so protein-coding genes - more than 98% of the genetic sequence inside each of our cells - was written off as "junk" DNA. Already falling out of favour in recent years, this concept will now, with Encode's work, be consigned to the history books.*" [15]

It was **The National Human Genome Research Institute** who launched a public research consortium named ENCODE, the Encyclopedia Of DNA Elements, in September 2003. It's mandate was to carry out a project to identify all functional elements in the human genome sequence, including Junk DNA. In 2012, thirty papers in the journals **Nature, Science, Genome Biology** and **Genome Research** published the results of the five-year Encode project. What did the results show>

> "*Encode is the largest single update to the data from the human genome since its final draft was published in 2003 and the first systematic attempt to work out what the DNA outside protein-coding genes does. The researchers found that it is far from useless: within these regions they have identified more than 10,000 new "genes" that code for components that control how the more familiar protein-coding genes work. Up to 18% of our DNA sequence is involved in regulating the less than 2% of the DNA that codes for proteins. In total, Encode scientists say, about 80% of the DNA sequence can be assigned some sort of biochemical function.*" [16]

What all this means that one cannot simply take a sample of the human genome and compare it with that of the genome a chimpanzee and then declare that they are almost 99% the same. DNA sequencing is far too complicated for such generalisation. Otherwise, one could say that humans have also descended from a common ancestor of pigs and humans because tests so far suggest, using the same criteria as that has been use for chimpanzees, pig DNA is 94% compatible with that of humans.

What I am about to do now is to show you how evolutionists like Richard Dawkins have used false data to develop a fantasy story, a myth to promote their atheistic world view of evolution. As Hitler once said, "If you tell a big enough lie and tell it frequently enough, it will be believed." In this evolutionists have succeeded magnificently. Let's take a look at the real facts of the case and not the fiction that they have so cleverly expounded upon.

THE MAKING OF A MYTH

As stated above it was Richard Dawkins in 1986 who set the stage for the myth that chimpanzees and humans shared more than 99 per cent of our DNA. We will learn more about Dawkins in the next chapter, but as the reader now knows our genome was not mapped until 2003 and 94% of the chimpanzee genome was not completed until 2005. So where did Dawkins get his information that allowed him to make such a preposterous presumption that would have a huge impact in what people believed about humans and our supposed common ancestor thereafter? The answer to the question lies with a study carried out by Allan Wilson and Mary-Claire King of the University of California that was published in *Science* magazine in 1975. [18]

In 1975, genetic engineering was still in its infancy, while at the same time the theory of evolution was in crisis, although reading books such as that of Dawkins would not have indicated as such. As Michel Morange, Professor of biology at the University Paris writing in the Journal of Biosciences relates:

> "*King and Wilson's publication was one of many in the 1960s and 1970s to provide molecular data on evolution and **to challenge traditional evolutionary models...The dominant role of natural selection was questioned by the neutralist theory, and also by an emphasis on the existence of constraints in evolution, as affirmed by Stephen Jay Gould and Richard Lewontin (Gould and Lewontin 1979). The model of punctuated equilibria - the alternation of stasis and rapid change - proposed by Niles Eldredge and Stephen Jay Gould (Gould and***

Eldredge 1977) was confirmed by the careful studies of Williamson (Williamson 1981). There was a huge debate to appreciate the significance of these observations, whether they demonstrated that the action of natural selection was limited by constraints in the construction of organisms." [19] [bold mine]

As to the study, Mary-Claire King described in a talk show how the 99% figure had been obtained. This is what she said:

"*So what we did in these days, this was by now the early 1970's - 1970, 1971, 1972 - was taken blood samples from lots of chimps who were living in zoos roundabout or who had been, who were visitable by people who were working in the field. And lots of blood samples from people roundabout. **And we compared them with the techniques of the time**. And we've showed using the techniques of the time that humans and chimpanzees share 98/99% of our genetic material.*" [20] [bold mine]

As Michel Morange rightly says, in 1975, genetic engineering was still in its infancy. For King and Wilson to claim that that chimpanzees shared 99% of our genetic material was grossly misleading as the genome would not be fully mapped for another thirty years after they made the announcement.

What King and Wilson really did was to compare a portion of DNA from both humans and chimpanzees, find a match and thereby assume that the rest of the DNA matched too. Using an analogy, it was like comparing the front wheel of a London bus and that of a front wheel of a car, which of course the construction of both wheels are almost identical, while failing to look at the big picture and observe that a bus and car are entirely different vehicles altogether. One could hardly describe the two vehicles as being only 1% different.

It is interesting to note what the concluding remarks of the 1975 King and Wilson's report says.

> *"The comparison of human and chimpanzee macromolecules leads to several inferences: I) Amino acid sequencing, immunological, and electrophoretic methods of protein comparison yield concordant estimates of genetic resemblance. These approaches all indicate that the average human polypeptide is more than 99 percent identical to its chimpanzee counterpart."* [21]

It is important to observe that King and Wilson do not claim that they measured the DNA genome of humans and chimpanzees. They simply said that the average human polypeptide was more than 99% identical to its chimpanzee counterpart, and therefore this infers that this is what they measured. In fact that is what they say, *"protein comparison yield concordant estimates of genetic resemblance."*

For the reader's information a polypeptide is a string of amino acids linked together, and such a chain might make up the entire primary structure of a simple protein. So we are talking about a protein that has been manufactured by the DNA and not the genome itself. Polypeptides are made when mRNA is translated, a process involving ribosomes, tRNA and amino acids to produce proteins.

That there will be a strong correlation between humans and chimps is understandable from the standpoint of intelligent design (not evolution), because proteins are the backbone of chemical machinery inside a cell. Cells have to have machinery for metabolism, for cell division, for translating DNA into proteins, for dealing with toxins, and for responding to the environment. The machinery has to accomplish many of the same things in cells of many kinds, so it should not be surprising to find that there are similarities among proteins not only between man and chimpanzee but throughout the world of living things. This is what King and Wilson really found.

The King and Wilson report also said, "*Although humans and chimpanzees have rather similar chromosome numbers, 46 and 48, respectively, the arrangement of genes on chimpanzee chromosomes differs from that on human chromosomes. Only a small proportion of the chromosomes have identical banding patterns in the two species.*" [21]

It is clear that the authors of the report were aware that their conclusion were tentative at best because there were issues, as aforementioned, that had not been addressed.

For Dawkins and other proponents of evolution, the study by King and Wilson seemed like Manna from heaven. They perceived that they had the "evidence" needed to target Creationists who by this time were making a comeback, and so they milked the study for all it's worth. Whether Dawkins and his evolutionist colleagues knowingly misrepresented the report to mislead people with false information or had innocently misinterpreted what King and Wilson said in their report is open to debate, but what is not debated is what they did with that misleading information. Through their network of pro-evolutionist journals and newspapers, Dawkins and his associates used their evolutionist propaganda machine to make sure that everyone knew about the "fact" that chimpanzees shared 99% of our DNA. This is how one of the greatest myths of the twentieth century was born and the rest, as they say, is history.

IF TRUTH BE TOLD

The first thing to keep in mind when genome mapping is done is that those carrying out that work are highly selective and only report the "best of the best" data. In many cases, pro-evolutionist scientists search only the protein coding gene sequences of preselected highly similar DNA which are known to be present in both species, which virtually guaranteed high levels of similarity. In other words 98% of the genome consisting of junk DNA is ignored and only the 2% of protein making DNA is used.

Doubts about what was being generally touted that the 98.5% (a compromise figure based upon King's 98/99%) claim for human and chimpanzee compatibility started to emerge from 2002 onwards. For example, *New Science* magazine published an article under the heading **Human-chimp DNA difference trebled** and said that, "*It has long been held that we share 98.5 per cent of our genetic material with our closest relatives. That now appears to be wrong. In fact, we share less than 95 per cent of our genetic material, a three-fold increase in the variation between us and chimps.*" [22]

The new 95% value came to light when professor Roy Britten (1919-2012), a molecular biologist based at the California Institute of Technology, began to question the methods used to arrive at the 98.5% figure that was being touted about. He recognised that the technique that had been used to obtain that figure was called a single base substitution, whenever a single "letter" base differs in corresponding strands of DNA from the two species.

Britten was famous for his discovery of repeated DNA sequences in genomes, and from his extensive knowledge about the subject he knew that there were also two other major types of variation that the previous analyses ignored. These were "insertions" and "deletions". "Insertions" occur whenever a whole section of DNA appears in one species but not in the corresponding strand of the other. Likewise, "deletions" mean that a piece of DNA is missing from one species. Together, they are termed "indels", and Britten seized the opportunity to evaluate the true variation between the two species when stretches of chimp DNA was published on the internet by teams from the Baylor College of Medicine in Houston, Texas, and from the University of Oklahoma.

When Britten compared five stretches of chimp DNA with the corresponding pieces of human DNA, he found that the DNA of both species was littered with indels. His comparisons revealed that the indels added around another 4 per cent to the genetic differences. This would bring the so called DNA compatibility between humans and chimpanzees to about 95%. Britten was not alone with his analysis. In 2006 *Scientific American* published an article titled, **Human-Chimp Gene Gap Widens from Tally of Duplicate Genes**, described the research headed by geneticist Matthew Hahn of Indiana University, Bloomington. The journal said, "*Researchers studying changes in the number of copies of genes in the two species found that their mix of genes is only 94 percent identical. The 6 percent difference is considerably larger than the commonly cited figure of 1.5 percent.*" [22]

The new figure of 94%/95% for human and chimpanzee DNA compatibility was beginning to cause some consternation in evolutionary circles, especially when early indications of the pig genome showed similar percentage compatibility between humans and pigs with the new figure. *The Swine Genome Sequencing Consortium (SGSC)* was formed in September 2003 by academic, government and industry representatives to provide international coordination for sequencing the pig genome. In a conference review report by a team of scientists headed by professor Lawrence B. Schook, Institute for Genomic Biology, University of Illinois, Urbana, IL, USA, in 2005 it was made clear that, "*The pig genome is of similar size, complexity and chromosomal organisation (2n = 38, including meta - and acrocentric chromosomes) as the human genome.*" [23]

Evolutionists therefore had reasons for concern and it was not unfounded. For example, when it was learned that junk DNA was not junk after all, Dawkins ran a damage risk exercise to try to distance himself from his previous outbursts of evolutionist piety.

Dawkins, 2009: on "junkDNA"

"it's full of junk, which is just as Darwinism predicted... how embarrassing for those creationists who say it shouldn't be!" (The Greatest Show on Earth, pp. 332-333)

Dawkins, 2012: on "junkDNA"

"*it's not full of junk, which is just as Darwinism predicted... nothing for the creationists to take advantage of here, move along!*" (Debate with Rabbi Sacks)

Evolutionists like Dawkins are now fighting with their backs to the wall. All of a sudden other scientists are now discovering that DNA exhibits evidence for intelligent design and are speaking out about it.

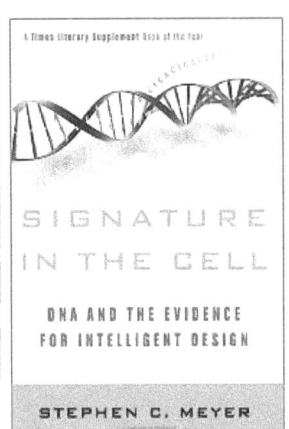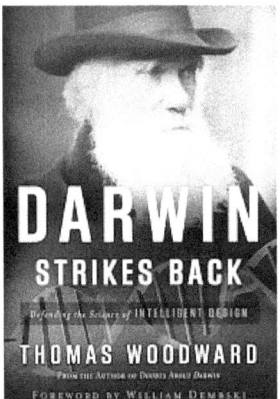

Icarus is a premier scientific journal published under the auspices of the American Astronomical Society's Division for Planetary Sciences (DPS) and contains articles that discuss the results of new research on astronomy, geology, meteorology, physics, chemistry, biology, and other scientific aspects of the Solar System or extrasolar systems. So when evolutionists saw the headline "Scientists dumbstruck: signs of intelligent design in DNA code", in the 14th March, 2013 edition of *Icarus*, they must have had a massive heart attack.

The researchers in the report believed they had found a message hidden in the DNA code of all living organisms that cannot be explained by chance, and which strongly adheres to the principles of mathematics and human language concepts, displaying "readily recognisable hallmarks of artificiality". The article ended with the words, *"The identity of any designer - whether alien or supernatural - remains unknown, but the study is groundbreaking in its implications."*

This news could hardly come at a worst time for evolutionists. With junk DNA now shown to perform essential functions, the ramifications of which could impact on their much promoted 95% compatibility between human and chimpanzee DNA, the evolutionist community had become dumb struck as another one of their long held beliefs had been disproved. This was the belief that the two sex-chromosomes would be nearly identical in humans and chimpanzees. It turned out that the chimp Y chromosome has only two-thirds as many distinct genes or gene families as the human Y chromosome and only 47% as many protein-coding elements as humans. Also, more than 30% of the chimp Y chromosome lacks an alignable counterpart on the human Y chromosome and vice versa. Let us take a closer look at this unexpected paradox and it's ramifications to evolutionary theory.

THE X AND Y CHROMOSONE PARADOX

autosomes   sex chromosomes

Chromosomes in humans can be divided into two types: autosomes and sex chromosomes. Certain genetic traits are linked to a person's sex and are passed on through the sex chromosomes. The autosomes contain the rest of the genetic hereditary information. All act in the same way during cell division. In humans, each cell normally contains 23 pairs (karyotypes) of chromosomes, for a total of 46 per cell. Twenty-two of these pairs, called autosomes are numbered by size and look the same in both males and females. The 23rd pair, the sex chromosomes, differ between males and females. Females have two copies of the X chromosome, while males have one X and one Y chromosome. Human cells have a total of 46 chromosomes per cell, apes like chimpanzees, bonobos and gorillas have 48 while pigs have 38.

By 2012 a number of well qualified geneticists were beginning to question the status quo. Among these was David Page, director of the Whitehead Institute of the Massachusetts Institute of Technology in Cambridge. There was an issue that he wanted to test. For almost a century, researchers thought that the Y chromosome, with far fewer genes than the X, was decaying. This was based upon the assumption that the Y chromosome has steadily lost genes as well as its ability to recombine and swap genes with the X chromosome. This suggested that the Y chromosome has long been isolated with just a couple of hundred genes, at most, whose job is to produce sperm and determine the sex of offspring. As a result, pro-evolution scientists predicted that the Y chromosome should be nearly identical in humans and chimpanzees, like the rest of the genome. They were wrong. Page and his team had just found that the Y chromosomes of chimps and humans are "*horrendously different from each other*". [24]

Page and his team found that the chimp Y chromosome has only two-thirds as many distinct genes or gene families as the human Y chromosome and only 47% as many protein-coding elements as humans. Also, more than 30% of the chimp Y chromosome lacked an alignable counterpart on the human Y chromosome and vice versa. This was totally unexpected and Page could not help express his astonishment, "*the relationship between the human and chimp Y chromosomes has been blown to pieces*".

This did not mean that Page and his team had abandoned their belief in evolution, it simply meant that they had to find another explanation from an evolutionary point of view to explain the strange discrepancy. So they came up with the idea that the rapid evolution and wholesale remodelling of the Y chromosome in both species had been caused by several mechanisms, including the competitive advantage gained by developing new genes for sperm production. [25] They had conveniently forgotten that for decades, the evolutionist paradigm had been that human and chimpanzee sex-chromosones would be virtually the same, and now it was found that this was completely erroneous.

> "*Just when we thought we were getting the sense that we had a pretty good picture of what our genome is like and how it evolved, we get tossed this curve ball,*" says geneticist Huntington Willard of Duke University in Durham, North Carolina. "*There are gems still buried in genomes that we haven't fully uncovered yet.*" [26]

"Holy crap!" exclaimed evolutionist John Hawks on his website.

> "*Indeed, at 6 million years of separation, the difference in MSY gene content in chimpanzee and human is more comparable to the difference in autosomal gene content in chicken and human, at 310 million years of separation. So much for 98 percent. Let me just repeat part of that: humans and chimpanzees, 'comparable to the difference ... in chicken and human'*". [27]

Hawks was referring to the Page's study in *Nature* which had the heading, **Chimpanzee and human Y chromosomes are remarkably divergent in structure and gene content**, which speaks for itself. Hawks, was amazed how little response there had been from his evolutionist colleagues. He exclaimed, "I can't believe how sedated the reaction to this paper has been so far. The outcome of the sequencing is really, really weird. More than thirty percent of the chimpanzee Y chromosome has no homolog in humans, and likewise for the human Y in chimpanzees."

This latest bombshell added to the ENCODE revelations has knocked the evolutionist community completely off balance. Long cherished beliefs have blown away so it is little wonder that response to the new data has been rather muted. What all this boils down to is that the genome is far from being understood and generalisations like the genome humans and chimpanzees (and pigs) are 94%/95% compatible is really untenable. Besides, as the sex chromosomes are also part of the genome, they should be included in the figure. When they are, "*Genome-wide, only 70% of the chimpanzee DNA was similar to human under the most optimal sequence-slice conditions.*" This figure was calculated by Jeffrey Tomkins, a Ph.D. in genetics from Clemson University who has 56 publications in peer reviewed scientific journals and seven book chapters in scientific books.

In 2013 Tomkins in conjunction with Dr Jerry Bergman with over 800 publications in 12 languages and 20 books and monographs under his name, published a comprehensive report detailing thirty separate experiments against four different high-quality human genome assemblies (GRCH37, GRCH36, Alternate SNP Assembly, and the Celera Assembly). The chimpanzee chromosomes, were sliced into new individual query files of varying string lengths and then queried against their human chromosome homolog using the BLASTN algorithm. Using this approach, queries could be optimised for each chromosome irrespective of gene/feature linear order. Non-DNA letters (gap filling 'N's) were stripped from the query data and excluded from the analyses.

The definition of similarity for each chromosome was the amount (percentage) of optimally aligned chimp DNA. The result was:

*"**For the chimp autosomes, the amount of optimally aligned DNA sequence provided similarities between 66 and 76%, depending on the chromosome**. In general, the smaller and more gene-dense the chromosomes, the higher the DNA similarity-although there were several notable exceptions defying this trend. Only 69% of the chimpanzee X chromosome was similar to human and only 43% of the Y chromosome. **Genome-wide, only 70% of the chimpanzee***

DNA was similar to human under the most optimal sequence-slice conditions." [28] [bold mine]

Although uninformed evolutionists and their disciples continue to espouse that human and chimpanzee DNA is 99% compatible, the true facts of the matter say otherwise. With what we now know can researchers in all honesty combine what's known and come up with a precise percentage differences between humans and chimpanzees? "I don't think there's any way to calculate a number", says geneticist Svante Paablo, a chimp consortium member based at the Max Planck Institute for Evolutionary Anthropology in Leipzig, Germany. [29]

CHROMOSOME 2 FUSION THEORY

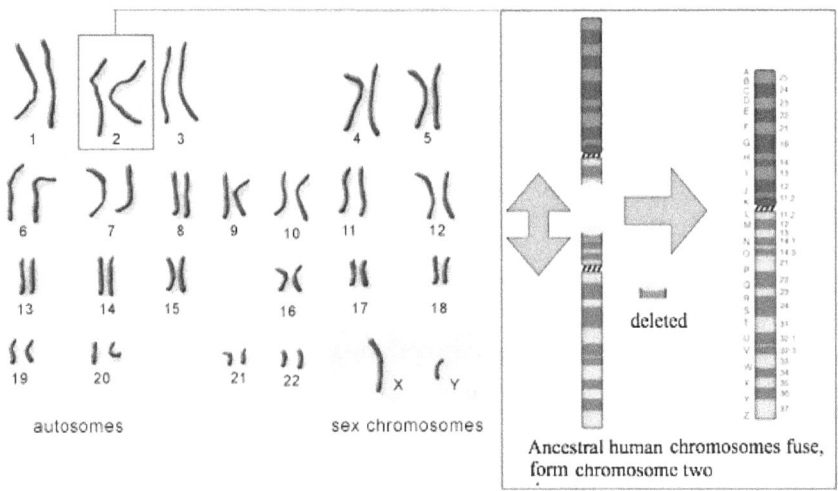

Ancestral human chromosomes fuse, form chromosome two

Earlier I mentioned that humans have 46 chromosomes and our so called ape relatives chimpanzees, bonobos, gorillas and orangutans have 48. Surely this must preclude Man having a common ancestor with these primates? One should never underestimate the ingenuity of evolutionists. They do have an answer. On the basis that we had descended from a common ancestor with the great apes, they reason that our ancestors must have lost a pair after our lineage branched off, some six million years ago. So they point to human chromosome 2 and claim that this chromosome fused (joined, merged) with the now missing pair (chimpanzee chromosomes 2A and 2B) to make one pair out of two. Voila! That explained the missing chromosome pair in humans and from henceforth it became known as the "chromosome 2 fusion" theory.

When chromosome 2 fusion theory was first postulated in 1982 by Jorge Yunis and Om Prakash at the Department of Medicine at the University of Minnesota Medical School, in their article **The origin of man: A chromosomal pictorial legacy** published in *Science* of that year, their presentation of DNA constructs did sound plausible.

> *"Evidence for a common ancestor of man and chimpanzee also comes from chromosome 2, since human chromosome 2 is more simply explained by telemeric fusion of a chimpanzee-like 2p chromosome and a 2q chromosome similar to that of chimpanzee or gorilla. These findings on chromosomes 2, 7, and 9 together suggest a common ancestor of human and chimpanzee. From this forefather, man emerged after the formation of a small pericentric inversion in chromosomes 1 and 18 and the fusion of chromosomes 2p and 2q to form the characteristic chromosome 2."* [20]

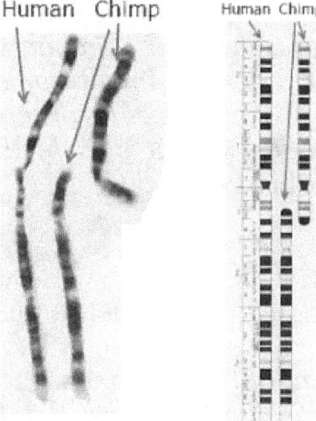

Looking at the DNA diagram in the Yunis and Prakash article, the DNA patterns do appear to support the evolutionist case at first glance because the human chromosome 2 does seem to line up with two chimpanzee chromosomes. But given that our chromosome structure is generally similar to that of apes anyway, it is not a stretch to assume that any 48 chromosome ancestor of modern humans might have also had a chromosomal scheme similar to that of apes, regardless of whether or not that individual was related to apes.

A major problem with the suggestion that the DNA sequence of the rest of human chromosome 2 closely matches very precisely the sequences of the two separate chimpanzee chromosomes, is fraught with difficulties. It is a claim unsupported by a lack of detailed comparative DNA sequence data, until recently. Earlier claims were based a chimpanzee rough-draft

DNA sequence assembly that was largely based on the human genome as a framework for its construction. When at last the first complete sequencing of the chimpanzee Y chromosome was made by Jennifer Hughes and colleagues, the chromosome sequences between chimpanzees and humans were so great that the heading of the report in *Nature* magazine in 2010 speaks volumes. It read, "**Chimpanzee and human Y chromosomes are remarkably divergent in structure and gene content.**" In other words the similarity of the DNA diagram in the aforementioned Yunis and Prakash article of 1982 that formed the basis for the fusion theory was in fact completely erroneous and misleading.

Anthropologist and evolutionist John Hawkins with his natural, sometimes comical style of writing describes the aforementioned report this way under the heading, **Unbelievable Y chromosome differences between humans and chimpanzees.**

> "*Holy crap! Indeed, at 6 million years of separation, the difference in MSY gene content in chimpanzee and human is more comparable to the difference in autosomal gene content in chicken and human, at 310 million years of separation. So much for 98 percent. Let me just repeat part of that: humans and chimpanzees, "comparable to the difference ... in chicken and human.*" [31]

Hawkins goes on to say:

> "*I can't believe how sedated the reaction to this paper has been so far. The outcome of the sequencing is really, really weird. More than thirty percent of the chimpanzee Y chromosome has no homolog in humans, and likewise for the human Y in chimpanzees.*" [32]

The sedated reaction that Hawkins refers is probably due to the shock that the evolution community has had to face when confronted by the substantial evidence presented by the report that gives the fusion theory a death blow. The report in *Nature* says:

> "*Previous models of Y-chromozone evolution treated the chromozone as uniform, homogenrous substrate for evolutionary change. In fact, the evolution of ampliconic sequences has outpaced that of X-degenerate sequences, and to such a degree that the ampliconic architecture of the common ancestor's MSY may be difficult to reconstruct even after an outgroup MSY sequence is mapped...*"

"About half of the chimpanzee ampliconic sequence has no homologous, alignable counterpart in the human MSY, and visa versa, compared to <10% of the X-degenerate sequence." [33]

For those who are not technical what the writers of the report was saying is that 50% of the Y-chromozone sequences do not match up. Put another way, there is no fusion sequence to be seen that fits the pattern that Yunis and Prakash had observed back in 1986, when only part of the chimpanzee genome had been mapped. One only has to look at the new data to see the huge differences between the chimpanzee and human Y-chromozone sequence.

Chimpanzee and human Y chromosomes are remarkably divergent in structure and gene content

Jennifer F. Hughes[1], Helen Skaletsky[1], Tatyana Pyntikova[1], Tina A. Graves[2], Saskia K. M. van Daalen[3], Patrick J. Minx[2], Robert S. Fulton[2], Sean D. McGrath[2], Devin P. Locke[2], Cynthia Friedman[4], Barbara J. Trask[4], Elaine R. Mardis[2], Wesley C. Warren[2], Sjoerd Repping[3], Steve Rozen[1], Richard K. Wilson[2] & David C. Page[1]

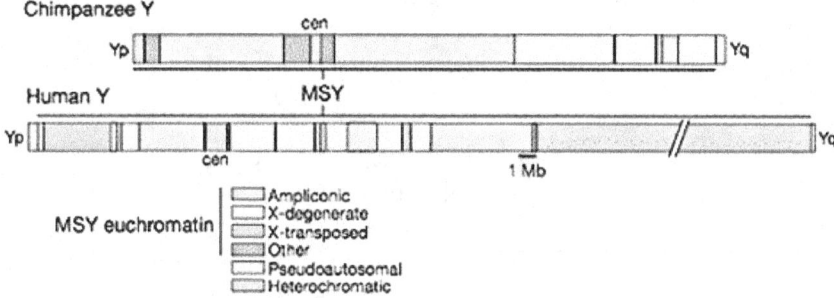

Comparison of chimpanzee and human Y chromes clearly showing large differences between the two as shown in the Jennifer Hughes report published in Nature 2010

nature Vol 463|28 January 2010|doi:10.1038/nature08700

Give Hawkins his due. Having been shown that the human Y chromosome differed in chimpanzees by as much as 30%, he refuses to let the evolutionist side down. He has after all spent most of his life supporting evolution and he is not going to give up easily and say that evolution had not taken place. Thus, he suggests that the main mechanism for the rapid structural evolution that the report suggests was probably autologous recombination, in other words derived from other DNA from the selfsame individual.

This he says, led to rapid structural evolution, but not necessarily any functional changes. This is of course is simply speculation and an attempt to dig the fusion theory out of a deep hole. He further suggests that widespread relocations of genes have a way of stripping them apart from upstream (or downstream) elements that may regulate their expression. But he now admits a tit-bit of information, that is most informative.

> "*Besides that, chimpanzees have lost several genes entirely, while humans have picked up a few that weren't in the common ancestor.*" [34]

This comes from the report where it says:

> "*In aggregate, the consequence of gene loss and gain in the chimpanzee and human lineages, respectively, is that the chimpanzee MSY contains only two-thirds as many distinct genes or gene families as the human MSY, and only half as many protein-coding transcription units.*" [35]

What is clear by the new report is that no matter how evolutionists struggle to interpret the new data, the so called banding pattern similarities that is said to provide evidence of common ancestry with apes is seriously flawed. But, even if for the sake of argument that the bandings were as Yunis and Prakash showed to be accurate, although we now know that they were inaccurate by at least 30%, could not the functional morphological and genetic similarities between humans and apes be just as easily explained as the result of common design?

What do I mean by common design? Let me explain by example. When Henry Ford built the first mass produced car, other manufacturers saw what he did and rather than develop something completely different from the ground up they used the same general template to build similar vehicles. As a result all vehicles have an engine, a gearbox, wheels, a steering mechanisms, body work that allowed the seating of passengers, the only difference is their final shape, colour, size and passenger capacity.

Would it not be reasonable to assume that an intelligent designer might have used a general template for the creation of the chimpanzee and bonobo, and then modified that design to create humans. Why reinvent the wheel? By merging chromosome 2 with what were two small chromosomes in chimpanzees, humans were endowed with significant beneficial enhancements not found in apes, such as bipedality and intelligence. Could this be why chromosome 2 is so different from that of the apes?

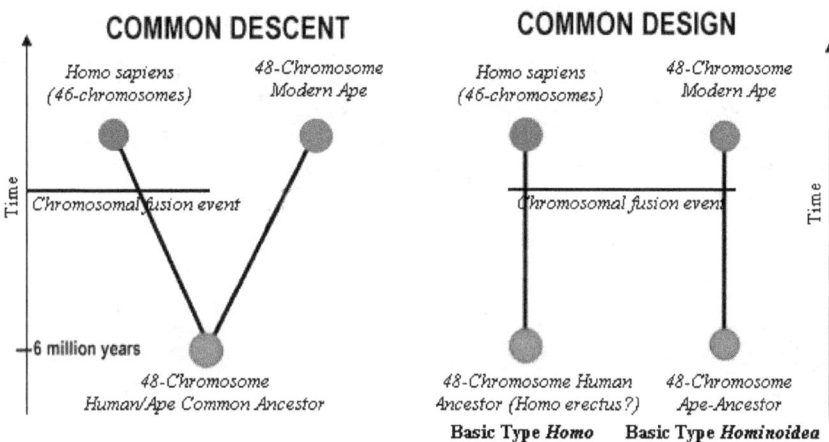

COMMON DESCENT

Homo sapiens (46-chromosomes)
48-Chromosome Modern Ape
Time
Chromosomal fusion event
6 million years
48-Chromosome Human/Ape Common Ancestor

COMMON DESIGN

Homo sapiens (46-chromosomes)
48-Chromosome Modern Ape
Time
Chromosomal fusion event
48-Chromosome Human Ancestor (Homo erectus?)
48-Chromosome Ape-Ancestor
Basic Type *Homo*   Basic Type *Hominoidea*

What do we know about chromosome 2? It is the second largest human chromosome, spanning more than 243 million base pairs and it represents almost 8% of the total DNA in cells. It therefore carries many genes some of which have been identified as being important in intellectual development, joint formation, hearing, facial structure and muscle development.

In conclusion, we do not know if this chromosomal fusion event happened on a line which leads back to some alleged common ancestor of apes and humans. All we do know is that this fusion event, if it had happen at all, could simply be a design attribute, a product of Intelligent Design happened in the line that led to you and me. Evidence of a chromosomal fusion event sometime in our past is by no means evidence that our line leads all the way back to apes.

WHAT DOES ALL THIS MEAN?

For a long time evolutionists have been playing the DNA game claiming that the so called 99% DNA compatibility between humans and chimpanzees proved a common ancestor. We now know that figure is greatly exaggerated and the real figure when taking in the sex chromosomes, indels and junk DNA into consideration a conservative estimate is around 70%. Rather than prove a common ancestor, all this does is show that humans are similar to other mammals, and nobody disputes this. Hence, when genomes of other animals are mapped, we should not be surprised to find similarities here too, and we do.

The **Genome Research** journal that focuses on research that provides insights into the genome biology of all organisms said:

"The mean percent identity of the alignments was highest for dog (79%), followed by cow (73.4%), primate (73.0%), and rodent (69%). Slight length discrepancies between the species imply that the primate-aligned regions are on average 0.5% longer than their cat counterpart, while those of rodent are 2.0% shorter, and cow and dog span regions of similar lengths." [36]

And we already know that pig DNA is 94% comparable to humans although I suspect that the sex chromosomes have not been taken into consideration in the figure, just as it wasn't originally with the chimpanzee genome.

This therefore begs the qustion. Why should evolutionists choose one set of figures such as that of chimpanzees and claim that this proves that we have evolved from a primate common ancestor when they ignore the rest? It should have been no surprise that living creatures on the earth possess very similar DNA structures and this is what genetics has been proved here. Living things' basic life processes are the same, and since human beings possess a living body, they cannot be expected to have a different DNA structures to other creatures.

Like other creatures, humans develop by consuming carbohydrates, lipids, and proteins, oxygen circulates through the blood in their bodies, and energy is produced every second in each of their cells by the use of this oxygen. For this reason, the fact that living things possess genetic similarities is no proof of the evolutionist claim that they evolved from a common ancestor.

In other words contrary to popular belief the science of genetics does not prove that humans developed from a common ancestor, nor does the fossil record for that matter. The enigma why humans are hairless and apes are not, continue to cause headaches for evolutionists and in last ditch effort and through desperation they have again turned to sexual selection for an answer.

CHAPTER 7
FROM THE ASHES OF THE PHOENIX

"Bateman's 1948 study is the most-cited experimental paper in sexual selection today because of its conclusions about how the number of mates influences fitness in males and females."
(Professor Patricia Adair Gowaty)

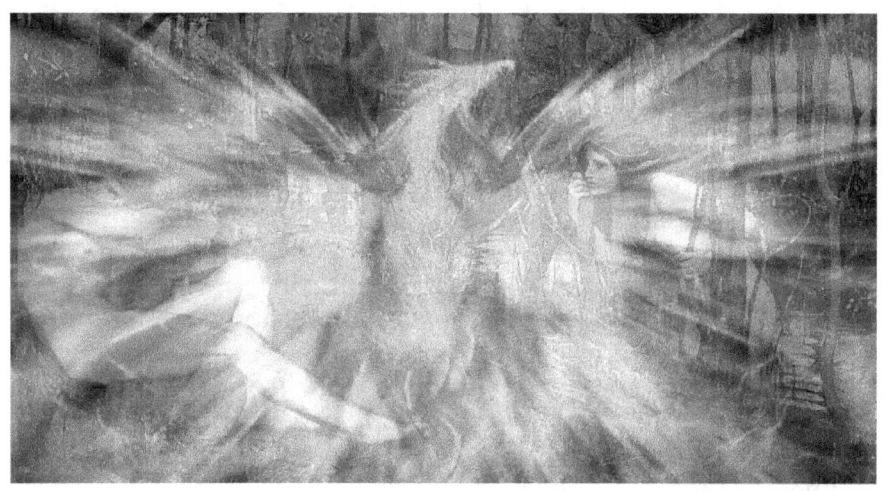

BATEMAN AND THE FRUIT FLIES | BATEMAN'S FLAWED EXPERIMENTS | BATTLE OF THE ATHEISTS | DAWKINS ON SEXUAL SELECTION | SEXUAL SELECTION PROVIDES NO ANSWER

It had taken Darwin thirty years to fully develop his second theory called sexual selection taken from his early notebooks to the 1871 book **The Descent of Man, and Selection in Relation to Sex**. For him it was the answer to all the problems that his earlier theory of natural selection had failed to answer, including the problem why Man had become hairless. However, most biologists of his generation and those that came after gave the theory a wide berth as professor Paulo Gama Mota of the university of Colombra, in Portugal notes.

"Darwin's theory of sexual selection is much less known that his theory of natural selection, and remained mostly ignored for

more than a century...The reception to the theory was cold. Although Darwin's contemporaries could accept that male-male competition does occur, and from it the evolution of weapons and size could follow, female choice was considered not acceptable." [1]

There were many leading biologists who rejected sexual selection including Thomas Hunt Morgan of whom I have already mentioned. However, it was Wallace, the co-discoverer of natural selection that swayed the majority of naturalists against sexual selection with sensible, solid and valid arguments. Writing in the Daily Mail newspaper in 2009 Wallace explained why Darwin's theory had been generally rejected.

"*As to his [Darwin's] conclusions on "Sexual Selection" there is equal diversity of opinion, two quite distinct phenomena being included under the term. These are (1) Weapons used in the combats of the males, which, being clearly useful to them and to the race, have been developed under the law of natural selection; and (2) colours and ornaments in the male sex only, the use of which is not so clear. It is Darwin's contention that the latter have been developed by the influence of female choice of the most highly ornamented males.* **The evidence collected by Darwin on this point is so abundant and so extremely interesting that most students were first carried away by it; but further consideration showed that direct evidence for any such choice was very scanty, while there was a fully equal amount against it.** *The fact that these colours and ornaments were almost equally developed in male butterflies, while it was almost impossible to postulate an identical æsthetic faculty among the females, together with our increasing knowledge of the various ways in which colour is of use, has led to a very general rejection of the theory of female choice.*" [2] [bold mine]

Wallace said that the weakness in Darwin's theory of sexual selection was that females have a very limited range of choice. Wallace argued that combat, or agility, or bodily vigour would have the greatest influence and that the part that remains to be played by ornament alone would be very small, even if it were proved, which it was not. Even so, Wallace was open to accept sexual selection if there was some tangible proof that supported the theory. He suggested experimental methods that could prove the theory. One of these was to cut off some tail feathers of pheasants and observe whether the hens rejected these males or not. Wallace said that until this was done, suppositions as to what determines the choice of the female could have but little value.

Wallace was a scientist in the true sense of the word. He wanted to test everything, and by suggesting an experiment to investigate sexual selection it shows his thorough approach to science and the scientific method. The scientific method involves two episodes. The first consists of formulating hypotheses; the second consists of experimentally testing them. What differentiates science from other knowledge is the second episode: subjecting hypotheses to empirical testing by observing whether or not predictions derived from a hypothesis are the case in relevant observations and experiments. A hypothesis is scientific only if it is consistent with some but not other possible states of affairs not yet observed, so that it is subject to the possibility of falsification by reference to experience.

Darwin had not done any testing to prove his theory about the peahens, and neither had anybody else for that matter. This is probably because by the time Wallace had written his arguments against the theory, most biologists had already agreed with him. Sexual selection just did not wash. As Zuleyma Tang-Martinez of the University of Missouri-St Louis, among others have said:

> *"While Darwin's theory of natural selection had been relatively widely accepted, his suggestion that a different process, sexual selection, could explain sexual dimorphism in animals had been, for the most part, met with scepticism. In fact, even Alfred Russel Wallace, the codiscoverer of the concept of natural selection, once wrote to Darwin and accused him of being 'un-Darwinian' because of the latter's advocacy of sexual selection."* [3]

It would be over a hundred years before the scientific method of testing along the lines that Wallace had suggested was finally carried out by three different research groups between 2007 and 2013. The results proved that Darwin had been wrong and that Wallace had been right. However, evolutionists had before then come up with a "proof" of sexual selection in the late 1940's and sexual selection once again appeared on the scene suggesting that Darwin had been right all along.

BATEMAN AND THE FRUIT FLIES

In 1948, English geneticist Angus John Bateman (1919-1996) published what became a very influential paper in the biological community. In that paper entitled "Intra-Sexual Selection in Drosophila" was published in the **Journal of Heredity** in which he reported his experiments with fruit flies. Using those experiments and referencing other scientists' observations, he concluded that, in general, males were promiscuous in their mating habits, while females were more choosy about their mates. "This rule", he said, "should be applicable to both animals and plants."

What Bateman did was to study small groups of equal numbers of male and female fruit flies (Drosophila melanogaster) that carried distinctive marker chromosomes. The flies were confined in small bottles and allowed to mate ad lib. He conducted six experimental series with 4-9 replications per series, 3- 5 fruit flies of each sex in each replicate, and with females allowed 3-4 days of laying. He then classified the offspring with regard to genotype (a group of organisms sharing a specific genetic constitution). From the experiments Bateman came to the conclusion that female reproductive success appeared to be limited by the number of eggs that could be produced, whereas for the male he suggested that fertility was seldom likely to be limited by sperm production but rather by the number of inseminations or the number of females available to him.

In layman's terms "Bateman's principle" as the experiments became to be called, said that females almost always invested more energy into producing offspring than males, and therefore in most species females were the limiting resource over which the other sex will compete. In other words, males competed with each other, and females become choosy in which males to mate with. As a result of being anisogamous, males were fundamentally promiscuous, and females were fundamentally selective.

These results appeared to prove sexual selection, just as Darwin had maintained. However, Bateman's experiments were seriously flawed and it would be decades before those flaws would be aired simply because the scientific method of repeating the tests had not been applied. When the experiments were repeated it was 2012 and found wanting, by which time thanks to Bateman, sexual selection had once again entered the mainstream of biological sciences and evolutionary thinking. The damage had been done and there was to be no turning back.

BATEMAN'S FLAWED EXPERIMENTS

In the intervening sixty years, Bateman's principle has been considered unquestionable truth in the evolutionary community and many scientists had gone so far as to call it a law. For example, in her book **Behavioral Mechanisms in Evolutionary Ecology** Emory University's Dr. Leslie Real writes:

> "One of the more persistent claims is that females will generally be more choosy than males in their selection of mates. Male fitness will thus be limited by access to females (leading to increased competition among males), while female fitness will

be limited by resources available for offspring production and development. This general claim has been elevated to the status of a law and often appears in the literature as 'Bateman's principle,' named after A. J. Bateman (1948)." [4]

There is only one problem with the aforementioned assessment. Bateman's principle is definitely not a general rule in nature, and more importantly, we now know that Bateman's original study was fundamentally flawed. Donald Dewsbury Professor Emeritus of Psychology at the University of Florida raises many issues that Bateman failed to take into consideration. In his paper **The Darwin-Bateman Paradigm in Historical Context** published in 2005 explains why Bateman's principle erred.

"It should be noted that all of Bateman's conclusions were based on differences in the representation of marker genes in progeny. He did not conduct systematic behavioral observations. Thus, as he noted, if a female mated with a male but no progeny resulted from that mating, the mating would have escaped detection. Further, Bateman could not distinguish single- and multiple- inseminations involving the same pair of flies. As he noted, "it might be possible when two matings occurred in quick succession for no progeny from the first mating to appear" (pp. 353-354). Thus, his estimates of the number of mates and matings must be regarded as minima. In fact, when Bateman wrote of the number of "mates," he actually measured the number of partners with which the individual produced offspring, not the number of individuals with which she mated or the number of times it did so." [5]

Why nobody attempted to repeat Bateman's tests as one would expect if the scientific method was applied is open to interpretation. Patricia Adair Gowaty, a distinguished professor of ecology and evolutionary biology at UCLA who repeated the experiments believes she has the answer.

"Our team repeated Bateman's experiment and found that what some accepted as bedrock may actually be quicksand. It is possible that Bateman's paper should never have been published... Repeating key studies is a tenet of science, which is why Bateman's methodology should have been retried as soon as it became important in the 1970s, she said...Our worldviews constrain our imaginations," Gowaty said. "For some people, Bateman's result was so comforting that it wasn't worth challenging. I think people just accepted it." [6]

Too late, the damage had been done. Since the 1950s Bateman's experiments had been widely cited in papers and text books for decades, and had been cited in nearly 2,000 other scientific studies as cast iron proof of Darwinist theory of sexual selection. [7]

Bateman's principle breathed new life into Darwin's second theory raising it out of the doldrums and by 1970 it had become firmly entrenched as an integral part of evolution. Pandora's box had been reopened and out flew Darwin's theory to infect the world once again, while all dissenting voices was cast aside and ignored. Today, professor Andres Paul of the Institute of Anthropology has made the following comment.

> "*After a long period of dormancy, Darwin's theory of sexual selection in general, and mate choice in particular, now represents one of the most active fields in evolutionary research.*" [8]

And, a survey on the ISI indexed journals indicates that sexual selection is increasing fast doubling the number of published articles per decade - 4500 in the 1990's and 10 000 in the 2000's. [9]

It was clear that Darwin's theory of sexual selection was here to stay, but just in Wallace's time, not all biologists were eager to jump ship and join the feeding frenzy. Michel Ohmer in her paper **Challenging Classic Sexual Selection Theory: The baby became the bathwater years ago, but no one noticed until now** writes:

> "*Based on the tenets that males ardently mate as much as possible and females are choosy and coy, essentially as a result of the mobility and abundance of sperm and immobility and scarcity of eggs, this deeply ingrained theory has formed the foundation of the study of animal reproductive behaviour for the past century. And yet, since its formation, not only has its basis been called into doubt more than once, the list of exceptions has only increased in length.*" [10]

Michel Ohmer has a BSc with Honours Ecology and Evolutionary Biology of Cornell University, New York, USA.

Another dissenting voice is that of Dr. Irwin Bernstein, Professor Emeritus of the Department of Psychology at the University of Costa Rica. His research has focused on the behaviour of monkeys and apes and his work has been published in over 200 journal articles, book chapters and other professional reports. Concerning sexual selection he sums up the problem with the theory.

"...it was much easier when I understood Darwin to say that macho males fight each other to claim females, and that the females choose the winners because these will be the best fathers for their sons. It does not seem quite that simple anymore... there are entirely too many animals that appear to violate that rule." [11]

In Wallace's day sexual selection as explained by Darwin could easily be refuted, but today the theory has been so modified to satisfy different criteria that it is difficult to find even one unifying definition of sexual selection. This means that not only is studying this theory complicated, arguing against it or one of its tenets is even more difficult.

Professor Tim Clutton-Brock of Cambridge University in his study **Sexual Selection in Males and Females**, published in *Science* 2007 makes this plain to see. He lists nine distinct but by no means exhaustive definitions of sexual selection, four of which involve sexual dimorphism and two of which only mention males, but all of which revolve around selection as a result of competition for mates. [12]

BATTLE OF THE ATHEISTS

Once again the battle lines are drawn as was the case in Wallace's day. Does sexual selection have any merit or not? That is the question that is being asked or argued about. Richard Dawkins for one has no doubts about the matter as we shall soon see. If you do not know who Dawkins is you must have been living on a different planet. He is an outspoken atheist, prominent critic of religion and 'high priest' of evolution.

Roger Downey writing in the magazine **Eastsideweek** in 1996 said:

"Evolution's first great advocate, 1860s biologist Thomas Henry Huxley, earned the nickname 'Darwin's bulldog' from his fellow Victorians. In our own less decorous day, Dawkins has earned an even stronger epithet: 'Darwin's Rottweiler.'"

For the purposes of this book it is important to learn a bit about Darwin's Rottweiler before we look at his explanation of human hairlessness through sexual selection. The reason why we need to do this will become abundantly clear as we discuss what he says about this subject. At first Dawkin's credentials appear impeccable suggesting that he knows what he is talking about. But the reality is that he is really a Jekyll & Hyde figure, a man who has two faces. One that appears to define him as a respectable scientist and the other that shows him to be a prejudicial despot with a chip on his shoulder.

Richard Dawkins is an English ethologist, evolutionary biologist, and a popular author who preaches to the masses about his atheist materialistic philosophy. He is an emeritus fellow of New College, Oxford, and was the University of Oxford's Professor for Public Understanding of Science from 1995 until 2008. He was listed by **Time Magazine** as one of the 100 most influential people in the world in 2007, and he was ranked 20th in **The Daily Telegraph's** 2007 list of 100 greatest living geniuses. And in a poll held by **Prospect** magazine in 2013, Dawkins was voted the world's top thinker based on 65 names chosen by a largely US and UK based expert panel.

From such venerable accolades it is clear that Dawkins has become a person of considerable renown but is this reputation due to his clarity of vision and devotion to evolution or is it for other reasons? Let us take a look and dig beneath the surface and see what skeletons we can uncover.

Dawkins is the man who declared, "*It is absolutely safe to say that if you meet somebody who claims not to believe in evolution, that person is ignorant, stupid or insane (or wicked, but I'd rather not consider that).*" [13] So when fellow atheist Antony Flew (1923-2010), a British philosopher and longtime emeritus professor of philosophy at the University of Reading declared that he now believed in God, Dawkins was stunned. What did Dawkins do? Instead of asking why Flew had turned and try to reason with him, Dawkins belittled and ridiculed the aged Flew with suggestions that he had lost his marbles. Flew responded by calling Dawkins a secularist bigot. [14]

For most of his life Flew had been a strong advocate of atheism, arguing that one should presuppose atheism until empirical evidence of a God surfaced. Such was his vocal and written attacks against those who advocated God, Flew has been called, "the world's most notorious atheist". As Mark Oppenheimer writing in the New York Times in 2007 says:

"Before the current crop of atheist crusader-authors - Richard Dawkins, Daniel Dennett, Christopher Hitchens - there was Antony Flew."

When the empirical evidence of God did finally surface, to his credit Flew kept true to his word and was prepared to accept it without reservation. What was that evidence? Flew said that his atheism was shaken loose by recent scientific discoveries, namely the evidence of the Big Bang, which assumes a beginning of the universe; DNA and the unlikelihood of a naturalistic explanation for its enormous complexity; and the lack of plausible theories explaining the first self-replicating forms of life. [15]

In his book **There Is a God: How the World's Most Notorious Atheist Changed His Mind.** that was published in 2006, Flew wrote:

"*I have followed the argument where it has led me. And it has led to accept the existence of a self-existent, immutable, immaterial, omnipotent, and omniscient Being.*" [bold mine]

Instead of investigating whether or not Flew's arguments had any merit Dawkins simply allowed his pride and prejudice to get in the way and he threw his then considerable influence behind the anti-God lobby to attack Flew with ridicule and derogatory remarks. Flew was prepared to admit when he had been wrong but not so Dawkins. Dawkins hated God and religion so much that he wrote in his book **The God Delusion**:

"[God is] a vindictive bloodthirsty ethnic cleanser, a misogynistic, homophobic racist, an infanticidal, genocidal, phillicidal, pestilential, megalomaniacal, sadomasochistic, capriciously malevolent bully."

With a mindset like that it is little wonder that Dawkins would not even consider the possibility of Intelligent Design, and anyone no matter who, will suffer his lashing tongue regardless of their esteemed position. Hence, when scientist and fellow atheist Lord Martin Rees was awarded the Templeton prize, Dawkins saw red and called Rees a "compliant Quisling" because of his reconciliatory views on religion.

Ironically, it was a prize that Dawkins might, possibly have won himself because it is awarded to scientists, philosophers, theologians, members of the clergy, philanthropists, writers, and reformers, for work that has ranged from the creation of new religious orders and social-spiritual movements to human sciences' scholarship, to research about the fundamental questions of existence, purpose and the origins of the universe. However, although the prize celebrates no particular faith tradition or notion of God, Dawkins saw yet another atheist turning away from "the faith".

Who is Lord Martin Rees? He is a scientist of considerable note. The **Guardian** newspaper compares the two atheist credentials.

> *"Richard Dawkins - author of "The God Delusion" and theorist of the selfish gene - could claim to be the most famous scientist in Britain. Lord Martin Rees - astronomer royal, former president of the Royal Society, master of Trinity College, Cambridge - is arguably the most distinguished."* [16]

Mark Vernon of the same newspaper described the different attitudes that Dawkins and Rees shared. He said:

> *"Dawkins and Rees differ markedly on the tone with which the debate between science and religion should be conducted. Dawkins devotes his talents and resources to challenging, questioning and mocking faith. Rees, on the other hand, though an atheist, values the legacy sustained by the church and other faith traditions. He confesses a liking for choral evening song in the chapel of Trinity College. It seems a modest indulgence. The ethereal voices of rehearsing choristers can literally be heard from his front door. But for Dawkins this makes the man a "fervent believer in belief". And that is a foul betrayal of science."* [17]

Vernon also said that awarding the Templeton prize to Rees suggested that science was beginning to reject the advocacy of the likes of Richard Dawkins. He could be right. **The Spectator** magazine in April 2013 made this statement.

> *"The atheist spring that began just over a decade ago is over, thank God. Richard Dawkins is now seen by many, even many non-believers, as a joke figure, shaking his fist at sky fairies. He's the Mary Whitehouse of our day."* [18]

The fact of the matter is that many people are just tired of Dawkin's platitudes against religion and have seen through his shallowness - and many of those people are scientists! In 2007, "Discovery Institute's Centre for Science and Culture...announced that over 700 scientists from around the world have now signed a statement expressing their scepticism about the contemporary theory of Darwinian evolution." In Dawkin's eyes those scientists, regardless of their qualifications and reputations are idiots. But he is not prepared to debate with any of them. He has too often been made to look a fool in such debates. In fact Dawkins has established a reputation for avoiding debates with his strongest opponents.

For example, **The Daily Telegraph** published a news story entitled: "Richard Dawkins accused of cowardice for refusing to debate existence of God" with Dr. William Lane Craig, one of the foremost apologists for Christian theism. Craig is a research Professor of Philosophy at Talbot School of Theology, in California, and the author of 30 books and hundreds of scholarly articles on Christianity.

A philosophy lecturer and fellow atheist, from Worcester College, Oxford by the name of Dr Daniel Came wrote to Dawkins urging him to reconsider his refusal to debate the existence of God with Professor Craig. In a letter to Dawkins Dr Came pleaded: *"The absence of a debate with the foremost apologist for Christian theism is a glaring omission on your CV and is of course apt to be interpreted as cowardice on your part."* [19] Needless to say no debate took place.

There are those who know why Dawkins refused to debate Craig. When it comes to theology, although he attacks religion passionately, it is well known that he knows little of the subject. Literary critic Terry Eagleton, theologian Alister McGrath, and science philosopher Michael Ruse all have accused Dawkins of being ignorant of theology and therefore unable to engage religion and faith intelligently.

On the few occasions that Dawkins has dared to debate noted theologians he invariably has come of the worse. It is little wonder therefore that he avoids such debates. Such was the case when he participated in a debate with the Reverend Giles Fraser, the former Canon Chancellor of St Paul's Cathedral and a vicar of the Church of England, during a BBC radio interview in 2012. The Telegraph among others reported the incident under the heading, **For once, Richard Dawkins is lost for words**, with a sub-heading, "***Atheists' arrogance is their Achilles' heel, as a cringemaking radio performance has proved.***" The conversation went this way. [20]

The two men were debating some new figures produced by Dawkins's think tank, the "Richard Dawkins Foundation for Reason and Science". The statistics purported to show that most people who identify themselves as Christian turn out, when questioned on what they actually think, to be overwhelmingly secular in their attitudes on issues ranging from gay rights to religion in public life. Dawkins's conclusion was that these self-identified Christians were "**not really Christian at all**". And to prove his point he announced, triumphantly, that an astonishing number of Christians couldn't even identify the first book in the New Testament. The exchange that next took place certainly made the arrogant Dawkins a lesson in eating humble pie.

Quietly, Fraser responded.

"Richard, if I said to you what is the full title of *The Origin Of Species*, I'm sure you could tell me that."

"Yes, I could," Dawkins responded, clearly indicating that he was ready for the challenge.

"Go on then," Fraser replied.

"'On The Origin Of Species' ... Uh. With, Oh God," Dawkins stumbled. "On The Origin Of Species.' There is a subtitle with respect to the preservation of favoured races in the struggle for life."

Fraser, of course, seized upon the opportunity to make his point that not being able to name a book doesn't necessarily have anything to do with one's deeply held beliefs or their convictions.

"You're the high pope of Darwinism," Fraser said. *"If you asked people who believed in evolution that question and you came back and said two percent got it right, it would be terribly easy for me to go and say that 'they don't believe it after all.' It's just not fair to ask people these questions. They self-identify as Christians and I think you should respect that."*

Here we find Dawkins stepping out of his usual area of expertise, biological evolution, and attempting to become atheism's greatest apologist. Unfortunately, like so many other atheists, he picks out easy targets to engage in order to avoid having to deal with any serious challenges to his beliefs. And when confronted by a more serious challenge, he finds out to his cost that he knows nothing at all.

Give him his due, Dawkins did try his luck with Lord Rowan Williams, former Archbishop of Canterbury and probably regretted it. Speaking at the Cambridge Union Society in January 2013 in opposition to the Dawkin's proposition that organised religion has no place in the 21st century, Williams was his usual calm, scholarly self. By the time the debate had finished, the vote showed that Dawkins had lost by 324 votes to 136. [21]

Understandably therefore, Dawkins infantile attacks on religion fails when confronted by those who he castigates. Nor has his distasteful antics gone unnoticed by many scientists. British theoretical physicist and emeritus professor at the University of Edinburgh Peter Higgs for example agrees with those who find Dawkins' approach to dealing with believers 'embarrassing'. Higgs says that what Dawkins does too often is to concentrate his attack on fundamentalists. But there are many believers who are just not fundamentalists. This is what he has to say in an interview with the Spanish newspaper *El Mundo*.

"Fundamentalism is another problem. I mean, Dawkins in a way is almost a fundamentalist himself, of another kind." [22]

Richard Dawkins has added scientist Dr. Stephen Meyer, author of **Signature in the Cell** and **Darwin's Doubt** to his list of people he refuses to debate. Stephen Meyer is a leading advocate of Intelligent Design, a theory that certain features of the universe and of living things are best explained by an intelligent cause, not an undirected process such as natural selection.

Meyer is a scientist using scientific methods to disprove evolution and religion does not come into the equation. This makes Meyer a dangerous advocate and so when Meyer formally challenged Dawkins to debate their contrasting views of evolution before the public, from a purely scientific viewpoint, Dawkins refused. He has done so ever since. Dawkins knows that in such a debate he would be roasted alive.

Although Dawkins has not worked in the science laboratory for decades, give him is due his books still have a widespread influence. They are a staple of atheists and humanists worldwide. He writes in a lucid style where he throws in just enough informative material from science to keep the reader believing that he knows what he is talking about, but this is mixed with many speculations, and not a few diversionary indulgences such as creationist-bashing, which many of his followers clearly like. But not all of them. On his blog atheist Mark Wallace says that Dawkins: an embarrassment to atheists.

> *"I hate, hate, hate the fact that this pompous, unpleasant, hectoring bully attaches his name ubiquitously to the belief that I happen to hold. Sometimes, it's almost enough to make me want to change back to believing in God - half to escape association with him, and half just to spite him."* [23]

And Martin Robbins of the **NewStatesman** writes:

> *"Increasingly, Richard Dawkins' public output resembles that of a man desperately grasping for attention and relevance in a maturing community."* [24]

Science is about the seeking of truth but it is clear from his writing and pronouncements that for Dawkins this is far from his mind. When we look at what he says it is clear that he has a specific agenda, a mission that transcends his scientific training. In his book **The God Delusion** published in 2006, Dawkins screams from his pedastal:

> *"I am attacking God, all gods, anything and everything supernatural, wherever and whenever they have been or will be invented."*

It is a well written book if you like to read a 400-page treatise of a man's hatred of God and religion. His followers loved it. The problem is that Dawkins has no real knowledge of the subject that he attacks with so much hostility - theology.

Terry Eagleton, Distinguished Professor of English Literature at Lancaster University; Professor of Cultural Theory at the National University of Ireland and Distinguished Visiting Professor of English Literature at The University of Notre Dame takes Dawkins to task with respects to his knowledge of theology. He describes Dawkins credentials to speak about theological issues thus:

> "*Imagine someone holding forth on biology whose only knowledge of the subject is the Book of British Birds, and you have a rough idea of what it feels like to read Richard Dawkins on theology. Card-carrying rationalists like Dawkins, who is the nearest thing to a professional atheist we have had since Bertrand Russell, are in one sense the least well-equipped to understand what they castigate, since they don't believe there is anything there to be understood, or at least anything worth understanding. This is why they invariably come up with vulgar caricatures of religious faith that would make a first-year theology student wince. The more they detest religion, the more ill-informed their criticisms of it tend to be.*" [25]

DAWKINS ON SEXUAL SELECTION

When people like Richard Dawkins use ridicule, profanities and abuse and not science to present their case, then clearly with nothing else to offer the writing is on the wall. But let us be magnanimous and just see whether his knowledge of Darwinian evolution can shed light on the mystery of Man's hairlessness. This subject was brought up in his book **The Ancestor's Tale: A Pilgrimage to the Dawn of Life** published in 2004 from which I quote freely for critical review as I am allowed to do.

In the book Dawkins begins the topic of sexual selection by saying, "The essential point is that male appearance and female taste evolve together in a kind of explosive chain reaction. Innovations in the consensus of female taste within a species, and corresponding changes in male appearance, are amplified in a run-away process which drives both of them in lockstep, further and further in one direction. There is no overweening reason for the one direction to be chosen: it just so happens to be a direction in which the evolutionary trend started. The ancestors of the peahens happened to take a step in the direction of preferring a larger fan. It kicked in and, within a very short time by evolutionary standards, peacocks were sprouting larger and more iridescent fans, and females couldn't get enough of them."

Here we see Dawkins uses the flowery language for which he is famous to put his point across but he writes in such away that the reader assumes that what he is saying is supported by evidence. But that evidence is sorely wanting as I have shown in chapter two of this book, and he speaks as if peahens have the same kind of tastes as humans. To say that "peacocks were sprouting larger and more iridescent fans, and females couldn't get enough of them" is ludicrous. Such words may endear him to his followers but what Dawkins says about the tastes of peahens have not been proven through the scientific method, unless Dawkins is a peahen in disguise and therefore can talk from experience. Now that is a thought!

Dawkins should have read the words of Thomas Morgan before writing the aforementioned paragraph in his book. Way back in 1894 Morgan asked whether we are to assume that those peahens whose taste has soared a little higher than that of the average select males to correspond, and thus the two continue heaping up the ornaments on one side and the appreciation of the ornaments on the other side? "No doubt an interesting fiction should be built up along these lines," he said, "but would anyone believe it, and if he did, could he prove it?" Dawkins had indeed built an interesting fiction along these lines and as we shall see, he cannot prove what he is saying. The trouble is his followers believes every word he says with almost religious fervour.

Dawkins continues his narrative it the book.

> "*Every species of bird of paradise, many other birds, and fish and frogs, beetles and lizards, zoomed off in their own evolutionary directions, all bright colours or weird shapes - but different bright colours, different weird shapes. What matters for our purpose is that sexual selection, according to sound mathematical theory, is apt to drive evolution to take off in arbitrary directions and push things to non-utilarian excess. The suggestion that arose in the chapters on human evolution that this is just what the sudden inflation of the brain looks like. So does the sudden loss of body hair, and the sudden take-off into bipedality?*"

Dawkins does not really say much here other than to express his view that animals zoomed off in their own evolutionary directions. See what I mean about flowery language? However, he does make reference to sexual selection in the aforementioned paragraph but we don't know which version of the theory he talking about? He does not tell us, not yet that is.

Dawkins also makes mention of "sound mathematical theory" to bolster his dialog on the subject. What is that? He does not say, but later in his book he refers to the work of Sir Ronald Fisher. Dawkins explains, "For the moment, we return to the simpler world of peacocks and peahens where females do the choosing and males strut around and aspire to be chosen. One version of the idea assumes that choice of mate (in this case choice by peahens) is arbitrary and whimsical compared with, for example, choice of food or choice of habitat. But you could reasonably ask why this should be. According to at least one influential theory of sexual selection, that of the great geneticist and statistician R. A. Fisher, there is very good reason."

You have to take Dawkins word on this matter because he next says in the book, "I have expounded the theory in detail in another book (The Blind Watchmaker, chapter 8) and I will not do so again." That book was published in 1996 and even in a very short time, Fisher's hypothesis founded upon Darwin's peahen fantasising and drooling over the tail feathers of the peacock has been shown through a number of recent experiments, to be completely wrong, as shown in chapter two of the present tome.

Who was Sir Ronald Fisher? Born in 1890 and living to 1962 he was an English statistician known for his important contributions to statistics, including his famous Analysis of Variance (ANOVA), a collection of statistical models used to analyse the differences between group means and their associated procedures (such as "variation" among and between groups). Fisher was an ardent promoter of eugenics, the study of methods of improving genetic qualities by selective breeding especially as applied to human mating.

Arguably Fisher's greatest work was that of his book *The Genetical Theory of Natural Selection* published in 1930 in which he worked out the mathematical probability of natural selection. It is because of this work that Dawkins called Fisher "*the greatest biologist since Darwin*" and this what Dawkins meant when he referred to the "sound mathematical theory" that drove evolution to take off in arbitrary directions in *The Ancestor's Tale*. There was just one problem that Dawkins failed to mention. In the opening sentence of Fisher's book, Fisher emphatically states that, as expressed in the title of his work, that **"NATURAL Selection is not Evolution."** [bold mine] Did you get that? Fisher's theory was based upon natural selection and not evolution.

Fisher clarifies what he means by that statement. "*Yet, ever since the two words [natural selection] have been in common use, the theory of*

Natural Selection has been employed as a convenient abbreviation for the theory of Evolution by means of Natural Selection, put forward by Darwin and Wallace." It was wrong equating evolution with natural selection as if they meant the same thing, which they do not. As he next says, "*This has had the unfortunate consequence that the theory of Natural Selection itself has scarcely ever, if ever, received separate consideration...The present book, with all the limitations of a first attempt,* **is at least an attempt to consider the theory of Natural Selection on its own merits.**" [26] [bold mine]

It is strange that Dawkins, the "high pope of Darwinism" as he has been called, did not highlight the fact that what Fisher was presenting was not based upon Darwinism (evolution) but on natural selection, which is something different altogether. Natural selection as I have discussed elsewhere is the same process as that of selective breeding (artificial selection) by Man, except it is nature that does the breeding through random processes and mutations. Hence, natural selection is not evolution but inherited variation within a particular species that allows that species to take on attributes such as colour, size and shape, as is the case of selective breeding. This is why Fisher wanted to write about the theory of natural selection based on its own merits, unlinked to evolution.

There is more. Dawkins also says in his book that innovations in the consensus of female taste are amplified in a run-away process. From this it is evident that he had in mind Fisher's "**runaway model**". In this Fisher describes a process which he calls "runaway sexual selection" to explain exaggerated characters of organisms such as the peacock's tail. It was a theory that was said to account for the rapid changes of specific physical traits in male animals of certain species but he made it clear that it did not apply to all species.

Fisher believed that selection of such traits were the result of sexual preference when members of the opposite sex found a particular trait desirable. This preference makes the trait advantageous, which in turn made having a preference for the trait advantageous. The process Fisher termed "runaway", because over time, it was said that it would facilitate the development of greater preference and more pronounced traits, until the costs of producing the trait balance the reproductive benefit of possessing it.

What Dawkins does not tell you is that, "*there is no empirical support for Fisher's hypothesis of runaway sexual selection.*" [27] The theory was dismissed by eminent biologists such as Wallace, Morgan, Huxley, and Lack, and that according to Malte Andersson, Professor emeritus, Animal Ecology at the University of Gothenburg:

> "*No critical test has been performed that supports Fisherian sexual selection and excludes the alternatives, or estimates their relative importance. The role of Fisherian sexual selection by mate choice in population divergence and speciation also remains to be clarified.*" [28]

Another thing that Dawkins fails to mention in his book is that Sir Ronald Fisher was a Christian and remained so for the rest of his life. As Jim Foley on his evolutionist website TalkOrigins admits:

"Sir Ronald Fisher - the most distinguished theoretical biologist in the history of evolutionary thought. He was also a Christian (a member of the Church of England) and a conservative whose social views were somewhere to the right of Louis XIV."

That being the case, one has to wonder why Dawkins speaks so highly of Fisher in view of the vitriolic hatred he has towards anyone else who has anything to do Christianity, and especially towards anyone who even hints at compromise such as Martin Rees as aforementioned. This smacks of a hidden agenda. What I mean by this is that Dawkins appears to call upon famous people whose reputations could give his pronouncements a semblance of respectability while hiding from his readers where the true loyalties of those people may really lie. I bet the reader of his book did not know that Fisher was a Christian, dare I say it, an Old Earth Creationist like myself.

Dawkins now gets to the heart of the matter and in *The Ancestor's Tale* he now calls upon Darwin to explain sexual selection. Dawkins writes, "Darwin's *Descent of Man* is largely devoted to sexual selection. His lengthy review of sexual selection in non-human animals prefaces his advocacy of sexual selection as the dominant force in the recent evolution of species. His treatment of human nakedness begins by dismissing - more glibly than his modern followers find comfortable - the possibility that we lost our hair for utilitarian reasons. His faith in sexual selection is reinforced by the observation that all the races, however hair or hairless, the women tend to be less hairy than men. Darwin believed that ancestral man found hairy women unattractive. Generations of men chose the most naked women as mates."

Here Dawkins passes the buck. He does not say that he believes in what Darwin says about sexual selection with respects to men preferring non-hairy men. Instead he simply outlines Darwin's theory, but in the process by inference he suggests that Darwin was right in his analysis. And to be doubly sure that the reader gets the message Dawkins next says:

"For Darwin, the preferences that drove sexual selection were taken for granted - given. Men just prefer smooth women, and that's that." [bold mine]

What an extraordinary thing to say; namely that, "**Men just prefer smooth women, and that's that.**" Did Darwin really say that? He did not say that in such colourful language but I suppose in around about way he did. And to put his point across and putting further words into Darwin's mouth Dawkins says:

"Nakedness in men was dragged along the evolutionary wake of nakedness in women, but never quite caught up, which is why men remain hairier than women."

Here we see a good example of poetic license at work with absolutely no evidence to support what Dawkins claims. In fact it is just a load of tosh and not worthy of a man of Dawkins' perceived stature. Then he digs deep and takes a dig at Wallace in his book.

"Alfred Russel Wallace, the co-discoverer of natural selection, hated the arbitrariness of Darwinian sexual selection. He wanted females to choose males not by whim but by merit. He wanted the bright feathers of peacocks and birds of paradise to be tokens of underlying fitness."

Here Dawkins reveals his prejudices. It was not a question of Wallace wanting this or wanting that. It was the second part of Darwin's theory that Wallace objected to as Fisher acknowledged.

"A. R. Wallace accepted without hesitation the influence of mutual combats of the males in the evolution of sex-limited weapons, but rejected altogether the element of female choice in the evolution of sex-limited ornaments."

What Wallace wanted to do was to see some tangible scientific proof to support what Darwin said in connection with female choice.

Being a world-class botanist and naturalist, Wallace had extensive experience in observing the mating habits of mammals and birds and there was nothing from which he observed that suggested females were any more sexually motivated to find a mate than males. As it turned out, Wallace was right to be cautious because three independent tests (Arita Takahashi, M., Arita, 2008, Roslyn Dakin, 2011; Jessica Yorzinsk, 2013) have since conclusively disproved Darwin's hypothesis of female choice.

Again Dawkins makes an outlandish statement. *"For Darwin, peahens choose peacocks simply because, in their eyes, they are pretty."* That is crazy talk by Dawkins. There is no evidence to support that human perception of beauty is to be found in any other creature other than in humans. Even Darwin admitted that *"Many will declare that it is utterly incredible that a female bird should be able to appreciate fine shading and exquisite patterns."* However, Darwin then proceeds to say with conviction that, *"It is undoubtedly a marvellous fact that she should possess this almost human degree of taste."*

What Darwin was saying was how marvellous it was that the peahen had the same human capacity for taste when in fact he had no evidence to support his claim. If he had such evidence, Wallace and other biologists would have had no difficulty endorsing sexual selection as a bona fide theory, but they rejected female choice outright.

Today, we now know that Darwin's female choice model is completely erroneous as we have seen with the recent aforementioned experiments. Peahens do not look at peacock display of feathers as being pretty as we humans do. To the peahen they indicate health not beauty, just as Wallace had suggested. As Dawkins says, "For Wallaceans, peahens choose peacocks not because they are pretty but because their bright feathers are a token of their underlying health and fitness." In this, Wallace was correct.

SEXUAL SELECTION PROVIDES NO ANSWER

Since the 1970s Darwin's sexual selection theory, once thought to have died a natural death has risen from the ashes and has seen a renaissance in some circles. It is even suggested that the theory has played a part in the evolution of music. Darwin had written in his **Descent of Man**:

> "... it appears probable that the progenitors of man, either the males or females or both sexes, before acquiring the power of expressing their mutual love in articulate language, endeavoured to charm each other with musical notes and rhythm." [29]

In fact Darwin devoted ten pages to bird song and six pages to human music, viewing both as outcomes of a sexual selection functioning as another kind of courtship display to attract mates.

Darwin's music hypothesis was expanded upon by Geoffrey F. Miller of the Centre for Economic Learning and Social Evolution University College in 2001. He says that:

> "*Music is a biological adaptation, universal within our species, distinct from other adaptations, and too complex to have arisen except through direct selection for some survival or reproductive benefit. Since there are no plausible survival benefits for music production, reproductive benefits seem worth a look. As Darwin emphasized, most complex, creative acoustic displays in nature are outcomes of sexual selection and function as courtship displays to attract sexual partners. The behavioural demographics of music production are just what we would expect for a sexually selected trait, with young males*

greatly over-represented in music-making. Music shows several features that could function as reliable indicators of fitness, health, and intelligence, and as aesthetic displays that excite our perceptual, cognitive, and emotional sensitivities." [30]

Professor Joseph Jordania, an Australian - Georgian ethnomusicologist, suggested that in explaining such human morphological and behavioural characteristics as singing, dancing, body painting, wearing of clothes, Darwin (and proponents of sexual selection) neglected another important evolutionary force, intimidation of predators and competitors with the ritualised forms of warning display. Warning display uses virtually the same arsenal of visual, audio, olfactory and behavioural features as sexual selection.

According to the principle of aposematism (warning display), to avoid costly physical violence and to replace violence with the ritualised forms of display, many animal species (including humans) use different forms of warning display: visual signals (contrastive body colours, eyespots, body ornaments, threat display and various postures to look bigger), audio signals (hissing, growling, group vocalisations, drumming on external objects), olfactory signals (producing strong body odours, particularly when excited or scared), behavioural signals (demonstratively slow walking, aggregation in large groups, aggressive display behaviour against predators and conspecific competitors). According to Jordania, most of these warning displays have been incorrectly attributed to the forces of sexual selection.[31]

For many people though the revitalised interest in sexual selection as a force in evolutionary biology, current in at least nine different formulations, is really nothing more that offering something to explain difficulties that cannot be answered in any other way. For example, Razib Khan, in his article in **Discover** magazine writes:

"Sexual selection is a big deal. A few years ago Geoffrey Miller wrote The Mating Mind: How Sexual Choice Shaped the Evolution of Human Nature, which seemed to herald a renaissance of the public awareness of this evolutionary phenomenon, triggered in part by debates over Amotz Zahavi's Handicap Principle in the 1970s. Of course Charles Darwin discussed the process in the 19th century, and it has always been part of the arsenal of the evolutionary biologist (I first encountered it in Jared Diamond's The Third Chimpanzee, where he lent some credence to Darwin's supposition that human racial differences may be a consequence of sexual selection). ***But this bump in recognition for sexual***

selection seems to be accompanied by its co-option as a deus ex machina for all sorts of unexplained events. And yet as they say, that which explains everything explains nothing."[bold mine] [32]

The word "deus ex machina" that Khan uses is defined in dictionaries as something or someone that comes in the nick of time to solve a difficulty, especially in works of fiction. Khan concludes his article by saying:

"*Rather than a specific answer to a given biological question sexual selection theory may be more useful as a way to explain the constant background flux of evolutionary process. At this point I am not convinced that it is robust enough to give us good "rough and ready" rules of thumb which we can apply as a sieve upon the welter of evolutionary genomic results.*" [33]

Although some biologists today have latched upon sexual selection to account for some peculiarities that they have seen in the animal kingdom, the arguments against the theory that relegated it to the doldrums of obscurity in the early part of the twentieth century remains as true today as it did then.

Wallace said in a letter to Darwin in 1868:

"*One difficulty to me is, that I do not see how the constant MINUTE variations, which are sufficient for Natural Selection to work with, could be SEXUALLY selected. We seem to require a series of bold and abrupt variations. How can we imagine that an inch in the tail of the peacock, or 1/4-inch in that of the Bird of Paradise, would be noticed and preferred by the female.*" [34]

That was three years before Darwin published his *Descent* that expounded upon his second theory, and it was a detail that was never answered.

Talking about natural selection Wallace later wrote:

"*The vast amount of the superiority of man to his nearest allies is what is so difficult to account for. His absolute erectness of posture, the completeness of his nudity, the harmonious perfection of his hands, the almost infinite capacities of his brain, constitute a series of correlated advances too great to be accounted for by the struggle for existence of an isolated group of apes in a limited area.*" [35]

Natural selection could not account for these issues and neither could sexual selection. Wallace said that the male against male fighting aspects were simply forms of natural selection, and that the notion of "female choice" was attributing the ability to judge standards of beauty to animals far too cognitively undeveloped to be capable of aesthetic feeling (such as beetles). Wallace also argued that Darwin too much favoured the bright colours of the male peacock as adaptive without realising that the "drab" peahen's colouration is itself adaptive, as camouflage.

There are so many problems with the Darwin's second theory that Thomas Hunt Morgan writing in 1908 said:

> "*In the light of the many difficulties that the theory of sexual selection meets with, I think we shall be justified in rejecting it as an explanation of the secondary sexual differences amongst animals...The theory meets with fatal objections at every turn.*" [36]

What all this boils down to is that sexual selection in whatever guise it has been resurrected in today still cannot account for Man's near hairless condition. Dawkins colourful rehashing of Darwin's theory of female choice is of no consequence because Darwin's main example of the peahens and peacocks sexual preferences have been resoundingly knocked on the head by recent experiments. Echoing the words of Morgan, sexual selection meets with fatal objections at every turn, even today.

Chapter 8
THE ENIGMA CONTINUES

*"Despite the title of Desmond Morris's 1967 book, "The Naked Ape",
scientists do not know when in evolutionary history the "great
denudation" took place. Or, for that matter, why."*
(The Economist, 18th December 2003)"

DARWIN'S SECOND THEORY DIES AGAIN | REDICULOUS THEORIES
ABOUND | DIVERTING ATTENTION ELSEWHERE | A PIG IN A POKE |
INTELLIGENT DESIGN | I CHALLENGE RICHARD DAWKINS

Richard Dawkins has made it clear in many of his writings that he
believes that the greatest driving force in evolution is that of natural
selection.

> *"Evolution by natural selection is the only workable theory ever
> proposed that is capable of explaining life, and it does so
> brilliantly"*, he says in the science journal *New Scientist* in 2005
> [1]

However, Dawkins' faith in the brilliance of natural selection is obviously
misplaced because the theory could not explain the virtually hairless
condition of humans, when it should have been able to do so. This

conundrum was acknowledged by both the two discoverers of the theory. Alfred Wallace said:

> *"Man's Naked Skin could not have been produced by Natural Selection."* [2]

At the same time Dawkins mentor Charles Darwin reluctantly had to agree with Wallace.

> *"No one supposes that the nakedness of the skin is any direct advantage to man;* **his body cannot have been divested of hair through natural selection**.*"* [3] [bold mine]

Knowing that natural selection could not account for Man's hairless condition Darwin needed to offer another solution to the conundrum and he was up for the challenge. In this he proposed his second theory, Sexual Selection and presented his case in a new book entitled, *The Descent of Man and Selection in Relation to Sex* which was published in 1874. In it he said:

> *"I am inclined to believe, as we shall see under sexual selection, that man, or rather primarily woman, became divested of hair for ornamental purposes; and according to this belief it is not surprising that man should differ so greatly in hairiness from all his lower brethren, for characters gained through sexual selection often differ in closely-related forms to an extraordinary degree."* [4]

SEXUAL SELECTION IS NOT THE ANSWER

Darwin's views on sexual selection were opposed strongly by Wallace as the private correspondence between the two naturalists testify. Every time Darwin resumed the debate with Wallace he always found himself at a disadvantage because the arguments that Wallace raised against his theory were very strong. Writing to Hooker on 21 May 1865, "I always distrust myself when I differ from him". Darwin tottered between acceptance and doubt of his new theory. On 16 September, he wrote to Wallace, "you will be pleased to hear that I am undergoing severe distress about the protection & sexual selection: this morning I oscillated with joy towards you: this evening I have swung back to old position."

In the end Darwin stuck to his guns and pronounced that female choice, as well as contests between males, were the driving forces of sexual selection. However, Wallace argued that male-male competition, while real, were simply forms of natural selection, and that the notion of "female choice" was attributing the ability to judge standards of beauty to animals far too cognitively undeveloped to be capable of aesthetic feeling (such as beetles). Wallace also argued that Darwin too much favoured the bright colours of the male peacock as adaptive without realizing that the "drab" peahen's colouration was itself adaptive, as camouflage.

So the two great naturalists argued and agreed to differ. Not that it mattered. Long after they had both died, sexual selection failed to get widespread acceptance. Thomas Hunt Morgan (1866-1945) was one of many leading scientists of the day and he rejected the theory outright. To be fair, Darwin did not have the technology to test out his sexual selection theory by the scientific method, but today such experiments have been made. As a result on numerous occasions these experiments of Darwin's "female" choice solution have failed to stack up to serious scientific scrutiny.

For example, the famous Bateman experiments in 1948 concerning the fruit flies, so often quoted in evolutionist text books as proof of the viability of sexual selection, is now recognised as being severely flawed. Gowaty, Kim, and Anderson (2013) recently replicated the study and showed that Bateman's methods were incapable of established the intended conclusion, which was that "sexual selection acted primarily on males through female choice and through male competition and profligacy in mating". The study said:

> *It seems that the modern bedrock of sexual selection may have been quicksand..."* and *"Given that Bateman did his study 60 years ago and given that its citation frequency soared in the mid-1970s, it remains curious that it was not repeated post-1972. We continue to wonder why.*" [5]

Gowaty and colleagues ask why Bateman's experiments were not subjected to further tests, and they suggest the reason was because of pre-existing cultural biases and that the dominating world view, meaning evolution, dampened any scepticism and objectivity. In other words, why question something that agreed with the prevailing established norm, namely evolution?

> *"Our field might profitably do some soul-searching: Why were Bateman's obvious errors overlooked for so long? As we said in our primary report, legions of graduate students have for the past 40 years read and discussed Bateman. Why did they not bring attention to the errors? Surely all of them, among biologists at least, understand the elements of mutation, inheritance and Mendelian genetics. Why did their professors not challenge Bateman's results? We are inclined to the idea that Bateman's results and conclusions are so similar to status*

quo, dominating world-views (competitive males, dependent females) that pre-existing cultural biases of readers may have dampened skepticism and objectivity. Perhaps lack of repetition is simply due to lack of professional incentives such as funding for repetitions." [6]

Michel Ohmer who received her Bachelor of Science with Honours from Cornell University in 2008 and currently working at the School of Biological Sciences of the University of Queensland is one of many scientists who question the viability of Darwin's sexual selection theory. In her article entitled, **Challenging Classic Sexual Selection Theory: The baby became the bathwater years ago, but no one noticed until now** she makes the following comment.

"A funny thing happens when an evolutionary biologist decides that one of Darwin's fundamental theories is inherently flawed: the scientific community defends itself as if mortally threatened... Bernstein (2003) sums up the dilemma we now face with our dearly beloved theory of sexual selection very well: "it was much easier when I understood Darwin to say that macho males fight each other to claim females, and that the females choose the winners because these will be the best fathers for their sons. It does not seem quite that simple anymore... there are entirely too many animals that appear to violate that rule"." [7]

In February 2006 Joan Roughgarden of Stanford University in California and two co-authors had a review paper published in *The Scientist* magazine which claimed that sexual selection theory, which emphasises the often-different interests of males and females, was fatally flawed. They argued that males were frequently characterised as competing to fertilise as many females as possible, while females seek relatively few, high quality mates, given their limited production of gametes. So the authors proposed an alternative perspective, in which they used cooperative game theory to set reproduction in an essentially mutualistic framework.

Reproductive partners enter into a "contract," Roughgarden told *The Scientist*, in which their best individual interests are served by working together to produce offspring.

The backlash to the Roughgarden's article was extraordinary. *The Scientist* magazine published an article in May 2006 entitled, **Sexual selection alternative slammed: Biologists write to Science to defend the theory of sexual selection**. In it forty biologists joined forces and contributed a total of ten letters to the science magazine all critiquing Roughgarden's review paper that had suggested that reproductive behaviour is explained better by cooperative game theory than by the theory of sexual selection first proposed by Darwin. In their response to the letters, Roughgarden and her colleagues refuted most of the criticisms, writing "*[our theory] is about the number of offspring successfully reared and is not an extension of sexual selection theory.*" *They add that "if sexual selection is correct, its credibility will be enhanced once it is successfully tested against alternative hypotheses.*" [8]

Whether one agrees with Roughgarden and her alternative theory or not is really relevant. The fact of the matter is that the majority of biologists do not accept Darwin's sexual selection theory as one that can be substantiated under scientific scrutiny. And, since 2006 when the forty biologists put pen to paper to attack Roughgarden, at least three rigorous science tests have been made which refutes Darwin's sexual selection as being a viable working theory. So what does one make of the way the forty biologists castigated Roughgarden when in fact they have since been proved to have been supporting a theory that has no credibility. I think Michel Ohmer puts things into perspective very well.

> "*In the many, many letters to the editor that Science received in response to Roughgarden et al. discrediting sexual selection, the number one attitude was "don't throw the baby out with the bathwater." (This very metaphor was actually used many times by many different researchers.) In other words, maybe sexual selection isn't all that bad; it just needs to be reworked a little. This stubborn clinging to a theory that has long been ingrained in our thinking should flash a warning in our heads. Since when do scientists pledge allegiance to theories? Are we not supposed to be discovering and explaining the reality of the natural world... fact not fiction?*" [9]

In other words, those who still doggedly uphold Darwin's sexual selection do so because they are not prepared to accept any evidence that is contrary to their pet theories. Darwin's theory is so ingrained into their way of thinking that to be told that they might be wrong is like telling religious fanatics who believed that the earth was flat, that the world was actually round.

As I said earlier, three significant and rigorous science tests have been carried since the biologists wrote their letters of protest. And their faith in Darwin's sexual selection has proved to be wrong. One of these tests were carried by biologist Mariko Takahashi of the University of Tokyo. It was a seven year study that finished in 2008 and the conclusion was that there was no evidence that peahens chose mates according to the quality of the peacocks' tails. Takahashi said that the findings were "at odds with Darwin's theory of sexual selection." Takahashi made three specific observations.

> *"Combined with previous results, our findings indicate that the peacock's train (1) is not the universal target of female choice, (2) shows small variance among males across populations and (3) based on current physiological knowledge, does not appear to reliably reflect the male condition."* [10]

The other two tests were carried out in 2011 by Roslyn Dakin from Queen's University in Kingston, Canada, and the other test two years later which was published in the 2013 edition of the *Journal of Experimental Biology, Positive Science* by Jessica Yorzinski. These tests showed that Darwin's sexual selection theory failed the scientific method.

It is a shame that it has taken so long to finally debunk a theory that had once been rejected for decades, then to be resurrected in the 1950's as a last ditch effort to explain many of evolution's anomalies, only to be discarded at last as a theory that has been found scientifically unsound.

REDICULOUS THEORIES ABOUND

Darwin's second theory of sexual selection, whether one accepts the theory or not, cannot account for human body hairlessness. To say that a female ape-like ancestor took a fancy to a male that had less hair than others of his kind is in the light of the mating habits observed of chimpanzees and bonobos a ridiculous suggestion. Then to suggest that their descendants did likewise so that eventually after several million year humans emerged in all his naked hairless glory is a fairy tale not worthy of a scientist of Darwin's reputation.

I suppose that as he and Wallace recognised that natural selection could not offer a solution, Darwin had to think up something that he hoped would prove right in the end. Unfortunately, sexual selection fails stringent science scrutiny and there are just too many animals that violate his female preferences rule. This has left biologists to pick up the pieces with a difficult problem to solve. How can they explain why evolution took upon itself to denude our human ancestors of their hair. The answer is they can't.

In her book *The Naked Darwinist* Elaine Morgan quotes the words of William Montagna, the famous biologist whose far-ranging studies of mammalian skin brought him the honour of being elected President of the

Society of Investigative Dermatology, previously restricted to M.D.s and later he became Director of the Oregon Regional Primate Research Centre. According to Morgan after years of research he and his team could not find a reason for human hair loss.

> *"William Montagna and his team of researchers, after years of intensive investigation into all aspects of ape and human skin, failed to arrive at any conclusion about the reason for our hair loss. He sadly concluded: "Since it is this single factor which constitutes the chief difference between the skin of humans and the skin of other mammals, we are left with the major objective of our study still unattained." This is an honest and tenable answer: "We don't know"."* [11]

Morgan rightly says that Darwin hazarded that our nakedness might be due to sexual selection. But apes are hairy animals, in which the signs of health and nubility are glossy coats, not sparse ones with bare patches. It seems unlikely, she says, that the males of one particular group of apes would break ranks and arbitrarily begin to yearn for females with bald bodies, any more than modern males hanker for bald headed women. [12]

It is worth noting that chimpanzees have beneath their fur light-coloured skin. Take away the fur and you've a light-coloured animal that, in the hot African sun, would be extremely vulnerable to the damaging rays of the sun. So you need dark skin. Thus, what advantage would there be for the random processes of evolutionary change be for the transition of our hairy ape-like ancestor to becoming hairless in the tropical heat, when that heat on light skin would be so damaging. None that's apparent. So which came first? Light to dark skin or the denudation of hair?

Paradoxes like this are common place when trying to explain what possible evolutionary advantage the loss of hair could bring to our supposed hairy common ancestor, which even Darwin acknowledged. This has led some Neo-Darwinists to propose some seemingly plausible theories, but also some ridiculous ones too. Professor Francis Ebling writing in the *Journal of Human Evolution* (January 1985) expresses what he feels about such theories.

> *"The evolution of near nakedness in the human species has been accounted for by a series of myths which owe more to the predilections of their creators than to the available evidence."* [13]

I have to agree with Ebling. Upon investigation, none of the theories that I have described in this book has been able to resolve the problem why humans are almost visibly hairless when our supposed ape cousins are not. Certainly, none have gained consensus even among evolutionists.

Having examined these theories I am reminded of a comment by comedian and television presenter Dara O Briain, who in my mind sums up the truth of the matter.

> *"Science knows it doesn't know everything; otherwise, it'd stop. But just because science doesn't know everything doesn't mean you can fill in the gaps with whatever fairy tale most appeals to you."*

DIVERTING ATTENTION ELSEWHERE

With no serious answers forthcoming, the problem of human hairlessness is understandably avoided altogether. Richard Dawkins for example simply shrugged his shoulders and said it was not important. Of course it is important because if evolution cannot explain our almost hairless condition, what can? Only Creation through Intelligent Design can offer a viable solution. But of course evolutionists are not going to admit this, so how are they going to convince the world that we have descended from hairy ape-like creatures? By deception of course. What would happen if it could be said that humans were biologically close to chimpanzees? That would certainly make a difference wouldn't it?

It so happens the science of genetics, which in the 1970's was going through a remarkable period of scientific research and coincidently this included the beginning of the mapping of the human and chimpanzee genome. Early reports came out that indeed that the biology of humans were so close to chimpanzees that there was only 1% difference between the human and chimpanzee genomes. For evolutionist theorists this was like manna from heaven. However, genome mapping was an infant science at this time and was based upon crude technology called DNA reassociation kinetics.

The results were in fact erroneous. Without understanding the technology used and not knowing how much of the genome had actually been checked, evolutionists like Richard Dawkins jumped at the 1% difference and announced to the world that chimpanzees shared 99% of human DNA. It was written in his book *The Blind Watchmaker: Why the Evidence of Evolution Reveals a Universe Without Design* published in 1986. In it he said:

> *"The last common ancestor of humans and chimps lived perhaps as recently as five million years ago, definitely more recently than the common ancestor of chimps and orang-utans, and perhaps 30 million years more recently than the common*

ancestor of chimps and monkeys. **Chimpanzees and we share more than 99 per cent of our genes."** [14] [bold mine]

Dawkins' announcement spread like a virus in the popular media and to the atheist community he was a hero. It helped him sell many books bringing him both fame and fortune. Before long every biologist under the sun was declaring the "fact" that chimpanzees and we share more than 99 per cent of our genes. It is still being touted even today by those who have not checked the evidence and who are unacquainted with the truth of the matter. It fact it would be another twenty years or more, after Dawkins book was published, when the human genome would actually be fully mapped, and the chimpanzee genome 99% done. So how could it be announced that 99% of human and chimpanzee DNA was similar when neither had been fully mapped?

Meanwhile, the evolutionary propaganda machine milked this erroneous fact for all its worth and Creationist bashing with this "fact" was very popular, and it still is. Dawkins often used it this way to target anyone who disagreed with him, proudly announcing that anybody not believing in evolution were idiots. But then the truth started to come out in dribs and drabs in scientific journals, which were mainly pro-evolutionist biassed. You were not going to see many headlines emphasising the fact that the "fact" was wrong.

When Dawkins made his famous statement only 2% of the DNA had actually been mapped. Yes! Only two percent!

The two percent was the area of the genome where the protein making DNA was located. The other 98% of the DNA that had not been mapped did not seem to do anything so it was called Noncoding DNA or junk DNA and was ignored. Consequently, when the entire human genome was finally mapped in 2003 and 90% of the chimpanzee mapped in 2005, it soon became clear that the claim of 99% compatibility was untenable. Dr. Todd Wood, an expert in genome comparison and former Director of Bioinformatics at the Clemson University Genomics Institute, did a BLAST analysis that indicated human and chimp DNA were roughly 95% similar.

Then it was learned that this figure was also similar to that of the pig DNA comparison of about 94%, measured in the same way. This was extremely embarrassing because the last thing evolutionists wanted the public to know was that, based upon their previous method of reasoning, man could have descended from pigs as well and surprise, surprise they were silent about this statistic.

In fact a recent study (2012), published in the journal *Nature* under the heading, **Pig geneticists go the whole hog** suggest that pigs are so biologically close to humans in so many ways that they have the 112 gene variants that might be involved in human diseases. The study said that knowledge of the pig genome is allowing scientists to try to engineer pigs that could be the source of organs, including heart and liver, for human patients. Pig organs are roughly the right size, and researchers hope to create transgenic pigs carrying genes that deceive the immune system of recipients into not rejecting the transplants. The study was led by scientists at The Roslin Institute at the University of Edinburgh and the Wellcome Trust Sanger Institute, Wageningen University and the University of Illinois. [15]

Genetically engineering pig DNA so that they could produce organs, including heart and liver, for human patients, is something that they cannot do with chimpanzees or bonobos. Why not if the DNA of chimpanzees and bonobos are only 99% different from humans? The fact of the matter is that the DNA between human and chimpanzees are by no means that close. Dr. Jeffrey P. Tomkins, former director of the Clemson University Genomics Institute, did a different BLAST analysis and concluded that the similarity was only 86-89% between humans and chimpanzees. Since then he and others have reported about 70% difference when taking into account the sex chromosomes, which discussed in the previous chapter.

> *"Only 69% of the chimpanzee X chromosome was similar to human and only 43% of the Y chromosome. Genome-wide, only 70% of the chimpanzee DNA was similar to human under the most optimal sequence-slice conditions. While, chimpanzees and humans share many localized protein-coding regions of high similarity, the overall extreme discontinuity between the two genomes defies evolutionary timescales and dogmatic presuppositions about a common ancestor."* [16]

What all his demonstrates is that one cannot make a declaration of such enormous proportions as saying that of chimpanzee and human DNA being 99% to 95% similar without consequences when it is discovered that similar DNA matching also indicate that pigs are also 94% similar. For example, Polish scientists led by professor Zdzislaw Smorag, reported in the *Warsaw Business Journal Poland* that:

"Pigs are very similar to humans in terms of their DNA makeup. Some 94 percent of pig DNA matches with the DNA of a human being." [17]

This study adds weight to what the Roslin Institute at the University of Edinburgh have been trying to do, aforementioned. In this case the researchers in Poland had successfully genetically modified four pigs so that they possessed a combination of two genes which made them genetically closer to humans than other pigs. This means, the scientists say, human immune systems would be less likely to reject the pigs' body parts and organs during a transplant operation.

According to some reports, baboon DNA is 91% similar to human DNA. [18] So when it was reported in *The Telegraph* (20 January 2015) under the heading, **Pig hearts could be transplanted into humans after baboon success**, one may understand how this was possible. In the study that was reported in the world news media, a genetically engineered pig heart was transplanted into a baboon and it survived more than a year without being rejected. This given leading scientists hope that animal parts could one day provide limitless sources of organs. "Pigs were chosen because their anatomy is compatible with humans" said Muhammad Mohiuddin of the Cardiothoracic Surgery Research Programme at the National Heart, Lung, and Blood Institute in the US. The study was presented at the 94th American Association for Thoracic Surgery annual meeting in Toronto.

In view of the successes made with research in pig tissue and organs for human transplants, and with an apparent 94% DNA compatibility to us, does that mean that we are more related to pigs than apes? Can you imagine what would have happened if Darwin declared that humans had descended from pigs! How does the saying go? "For they sow the wind, and they shall reap the whirlwind!" Those who had advocated the erroneous "fact" that chimpanzee and human 99% had similar DNA were about to see their egotistic arrogance come back to haunt them in a way that they never would have imagined.

A PIG IN A POKE

Evolutionists like Dawkins may have diverted attention away from the enigma of human hairlessness by declaring that chimpanzee and human were 99% similar but sooner or later, the truth would catch them out. But not before one of their number used the revised figure of 95% for chimpanzee and human similarity, together with the 94% resemblance of chimpanzee and pig DNA and establish a new theory. The evolutionist suggested that the reason for the almost hairless condition of human could be explained by porcine ancestry. Shock! Horror! This was even worse than Darwin's sexual selection model where sex is play a predominant factor in the theory. Now someone was suggesting that our common ancestor had sex with a pig and saying that we were the descendants of this unnatural union.

According to one evolutionist we lost our hair because one of our ape-like ancestors mated with a pig and as a result we are descended from this common ancestor.
Mmmmm! Perhaps there is more to this cartoon than meets the eye.

The person concerned with this theory is Dr. Eugene McCarthy, and he is widely recognised as a leading expert on hybridisation, the act of mixing different species or varieties of animals or plants to produce hybrids. One of the things he said to support his theory was that the elasticity of our skin was very unusual. Whereas the skin of the great apes and that of some simian primates have variable amounts of elastic fibres, in no animals, regardless of sex, age, or locality has it been found the abundance of elastic tissue characteristic of human skin. However, when comparing human skin with that of pigs, it is observed that one of the most striking resemblances between these two skins [pig and human] is the large content of elastic tissue in the dermis.

There is more evidence to support his theory which McCarthy explains:

"So, in the pig, we have a sparsely haired animal with a fatty, stretchy skin supplied by musculocutaneous arteries. The surface of the hairy skin is marked by congenital lines similar to those seen in human beings, and the patterning of the epidermal-dermal junction is also quite similar in the hairy skin regions."

"Under the hypothesis that we are considering, it makes little difference that pig skin differs from human skin in other ways. The essential point is that, in those cases in which our skin is peculiar for a primate, an explanation for each such anomaly can be found in the skin of pigs." [19]

This is not what Neo-Darwinists wanted to hear because what McCarthy was saying about human and pig skin was true. TP Sullivan of the University of Miami School of Medicine, Department of Dermatology, Miami, Florida 33101, USA with colleagues came to the same conclusion.

*"**Anatomically and physiologically, pig skin is more similar to human skin.** The many similarities between man and pig would lead one to believe that the pig should make an excellent animal model for human wound healing... Over 180 articles were utilized for this comparative review. Our conclusion is that the porcine model is an excellent tool for the evaluation of therapeutic agents destined for use in human wounds."* [20] [bold mine]

Then newspapers like the Daily Mail published stories like, **How Animal Farm was right: Pigs really ARE almost identical to humans, say scientists**, which reported on the findings, published in the journal *Nature* that show that pigs suffer from the same genetic and protein malfunctions that account for many human diseases, including Alzheimer's, Parkinson's and obesity. [21] **Doctors graft pig's skin onto burned child in Zigong** was another story recently published where Doctors in Zigong, China, successfully grafted pig's skin onto the body of a 10-year-old boy who suffered burns on over 70 percent of his body. [22]

McCarthy's theory is the evolutionist community's worst nightmare. They can't have their cake and eat it too! How can they possibly promote their theory that man had descended from a hairy ape-like common ancestor and offer DNA "proof" as cast iron evidence, when the same DNA methods showed that human and pig DNA was also very similar. Worse still, it is the tissue of pigs that is being used for human medical experiments and not derived from our "closest cousins" the great apes.

What did the evolutionists do when confronted with this paradox? They did what they always do when faced with this kind of dilemma. They distanced themselves from McCarthy calling him names that I cannot print here. But McCarthy has a Ph.D and an enviable track record for hybridisation. He was not going to be silenced easily. He responded:

"For the present, I ask the reader to reserve judgement concerning the plausibility of such a cross. I'm an expert on hybrids and I can assure you that our understanding of hybridisation at the molecular level is still far too vague to rule out the idea of a chimpanzee crossing with a nonprimate." [23]

Not everyone rejects McCarthy's theory out of hand. Dr Chris Millar, from Ballarat, Victoria, Australia wrote:

"As a clinician and scientist with medical training it is a joy to find a theory so carefully and elegantly presented. My interest in the hybrid nature of modern man led me to Eugene McCarthy's website and lifework. What a revelation! Surprising and shocking. Such is the nature of truth sometimes. Life will never be seen in the same way after reading this work." [24]

Stephen Garcia, a mechanical Design Engineer from Guanajuato, Mexico also wrote:

"Your conjecture is not unlike trying to reverse engineer a human being. Logically it all makes a good argument, down to the detailed level you've taken it to. I imagine that working with hybrids you HAVE to do that - even in cases where you may not think so. Logically your arguments make a lot of sense. And the corollaries and ramifications all seem to come true. I am impressed, frankly." [25]

Such positive comments of course are few and far between, yet McCarthy does present a case that is reminiscent of the evidence that was presented by Darwin and his followers in the nineteenth century that convinced people that humans had descended from hairy ape-like ancestors. Let me show you what I mean. John Hewitt who has a background in physics and neuroscience describes McCarthy and his theory in the following way.

"Dr. Eugene McCarthy is a Ph.D. geneticist who has made a career out of studying hybridization in animals. He now curates a biological information website called Macroevolution.net where he has amassed an impressive body of evidence suggesting that human origins can be best explained by hybridization between pigs and chimpanzees. **Extraordinary theories require extraordinary evidence and McCarthy does not disappoint. Rather than relying on genetic sequence comparisons, he instead offers extensive anatomical comparisons, each of which may be individually assailable, but startling when taken together...** *The role of hybridization in driving morphological*

change, as McCarthy has observed time and time again, particularly in his studies of avian species (Oxford University Press, 2006), may be the most powerful mechanism of all." [26]

Isn't this exactly how Darwin and other naturalists of his day established the link between apes and humans by comparing anatomical similarities? Early evolutionist books are full of anatomical comparisons which is offered as proof of our primate heritage. Thomas Huxley, for example, wrote in his book **Evidence as to Man's Place in Nature** published in 1863 the following comparisons between humans and apes.

"Is man so different from any of these apes that he must form an order by himself?" (p85). "It is quite certain that the ape which most nearly approaches man is either the Chimpanzee, or the Gorilla..." (p86). Thus, whatever system of organs be studied, the comparison of their modifications in the ape series leads to one and the same result - that the structural differences which separate Man from the Gorilla and the Chimpanzee are not so great as those which separate the Gorilla from the lower apes" (p123). "But if man be separated by no greater structural barrier from the brutes than they are from each other - then it seems to follow that... there would be no rational ground for doubting that man might have originated... by the gradual modification of a man-like ape." [27]

Interestingly, when John Hewitt followed up his article on McCarthy's theory reflecting the comments that the magazine received he noted that despite the critical and personal attacks against McCarthy, his critics could not offer any decent scientific arguments against the theory.

*"There was considerable fallout, both positive and negative, from our first story covering the radical pig-chimp hybrid theory put forth by Dr. Eugene McCarthy, a geneticist who's proposing that humans first arose from an ancient hybrid cross between pigs and chimpanzees. **But by and large, those coming out against the theory had surprisingly little science to offer in their sometimes personal attacks against McCarthy.** As many critics noted, the advancement of scientific knowledge does not require disproving every radical theory that comes along. **It seems, however, that decent arguments against the hybrid origins theory are surprisingly hard to find.**" [28] [bold mine]*

Undeterred by his critics McCarthy continues to elaborate upon his chimpanzee and pig hybridisation theory through his website. At the same he would no doubt have observed the periodic news items that appear in the media of pig organs and tissue for human medical conditions makimg newspaper headlines that seem to lend credence to his arguments. A growing number of scientists now recognise the similarity between human and pig and have used this knowledge to implement life saving research. It is strange that one never hears of chimpanzee or bonobo tissue being used for similar purposes, especially when they are supposed to be our closest cousins with their DNA described as being closest to ours than any other mammal.

Coming back to the issue of human hairlessness, McCarthy declares that our hairless skin and subcutaneous fat could be explained by the hybridisation between pigs and apes as a result of their mating. This resulted in the birth of the earliest hominid pig/ape hybrids millions of years ago and that subsequent mating within this hybrid swarm eventually led to the various hominid types and to hairless modern humans.

For the record, as an Old-Earth Creationist I do not subscribe to McCarthy's theory, but then I do not endorse the evolutionary position that man descended from ape-like ancestors either. Darwin and other naturalists of his generation used the same arguments and tactics that McCarthy has used and I reject evolution (not natural selection) for the very same reasons. Even so, McCarthy's explanation is as good as any that evolutionists have presented thus far with respects to our almost naked skin.

There is one theory that is far superior to any thus far presented by evolutionists, and the science of genetics provides the key to unravelling Darwin's Enigma and that is **Intelligent Design.**

INTELLIGENT DESIGN

By the turn of the 21st century things have been going from bad to worse for evolutionists and their atheistic naturalist theories. For example, when the two sex chromosomes were taken into consideration, the human to chimpanzee DNA compatibility dropped to about 70%. Furthermore, as improvements in DNA testing methods improved, it was discovered that junk DNA was not junk after all and performed vital functions that we are only now beginning to understand. What all this meant was that DNA was far more complicated that had been previously realised and that making comparisons between humans and other animals such as chimpanzees and pigs is really an invalid argument for proving evolution. All that the science of genetics has done is to show that humans are similar to other mammals, and nobody disputes this, not even Creationists.

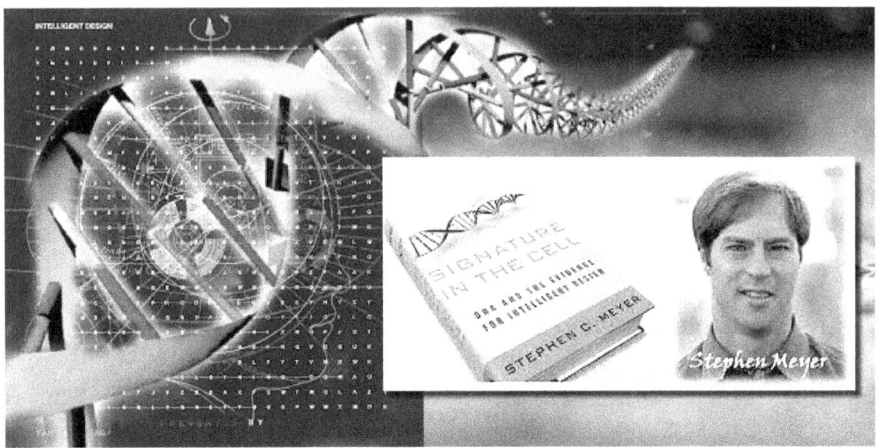

It should have been no surprise that all mammals on the earth possess very similar DNA protein making structures and this is all that investigations of the genome has been proved. Living things' basic life processes are the same, and since human beings possess a living body, they cannot be expected to have a different DNA structures to other creatures. The fact that living things possess genetic similarities is no proof of the evolutionist claim that we evolved from an ape-like common ancestor. In fact, Intelligent design is able to explain similarities in DNA far better than the evolution model. Designers often make different products by the utilisation of similar parts, materials and arrangements. The common percentage declared by evolutionists pertains to the regions of our DNA that result in proteins. It makes more sense if the data for the Designer of nature to have used the same proteins to perform the same functions in a variety of organisms.

There is one school of scientists that atheists like Dawkins fear the most and who refuses to debate, especially on matters of DNA because here, Dawkins is way out of his comfort zone. One of these ID scientists is Stephen Meyer the Cambridge-educated chemist and scientific historian who has puts forth his own case for Intelligent Design in his book **Signature in the Cell and the Evidence for Intelligent Design.** One reviewer of Meyer's book, Ignazio de Vega, wrote:

> "*In its calmly-reasoned 400 pages (with an extra 100 tightly-packed endnotes), Meyer constructs the strongest argument yet made for the theory of Intelligent Design, and he does it without once advocating any living God.*" [29]

Dawkins whose personal crusade to target anyone, Creationist or scientist who believes in God, is well documented, has with Meyers, a formidable adversary. Meyers does not say anything about God in his books but tackles the subject of DNA from a purely scientific and rational perspective. And, in the process destroys Dawkins arguments that DNA proves evolution.

> "Years ago, Richard Dawkins characterized our increasing knowledge of DNA as "the final, killing blow to the belief that living material is deeply distinct from nonliving material," but Meyer tackles this and other cautions head-on (his serial dismantling of Dawkins throughout the book is conducted with a very satisfyingly mandarin delicacy). Unlike many other proponents of Intelligent Design, Meyer isn't afraid of the newly-revealed intricacies of DNA: he welcomes them." [30]

The Signature in the cell explains in no uncertain terms why the genetic code would be impossible for nature to create by randomly combining primordial molecules. Instead, Meyer gives creditable proof of design, intent and purpose in the architecture of cells. Meyer is absolutely thorough and relentless in his approach to this topic, and all his arguments are not only powerful in terms of their striking common-sense logic, but also couched in meticulous research of actual scientific discoveries and methods of science itself. It is little wander that Dawkins refuses to debate with him.

The issue I am raising here is this. Evolutionists like Dawkins are quite happy to make reference to the science of genetics, with particular emphasis on DNA of the human genome, to prop evolution theory and twist the facts accordingly, but when push comes to shove, when their "evidence" is found wanting, they somehow carry on as if what they have said is true regardless of the facts. Then they have the audacity to dispute powerful evidence that the information code behind DNA sequencing shows evidence of an intelligent agency being involved.

For example, Dawkins says in books such as **The Blind Watchmaker**, "Biology is the study of complicated things that give the appearance of having been designed for a purpose", then proceeds to argue that they were not. But when Ben Stein (in the recent documentary "Expelled - No Intelligence Allowed") asked him at the very end of the documentary, "What do you think is the possibility that Intelligent Design might turn out to be the answer to some issues in genetics, or evolution?" Dawkins responded by saying:

> "It could be that at some earlier time, somewhere in the universe, a civilization evolved by probably some kind of Darwinian means to a very, very high level of technology - and

*designed a form of life that they seeded onto perhaps this planet. ... **And I suppose it's possible that you might find evidence for that if you look at the details of biochemistry, molecular biology, you might find a signature of some sort of designer.*** " [31] [bold mine]

Talk about beating around the bush, but Dawkins is very canny and is playing it safe. Meyer and others have shown that there is evidence of a signature of intelligence behind the cell's structure, and so here Dawkins is keeping his options open. Even so, he is not relinquishing his core belief in evolution and on this he makes this clear by saying, "somewhere in the universe, a civilization evolved by probably some kind of Darwinian means to a very, very high level of technology - and designed a form of life that they seeded onto perhaps this planet".

Of course, he does not go into how the life forms that developed the hypothetical highly technical civilisation came in to being other that evolution, but presumably, as long as that God is not involved he is happy with this. As Stein said later, "So Professor Dawkins was not against intelligent design, just certain types of designers, such as God." Now the clincher!

Dawkins admits to the possibility that these beings "designed a form of life that they seeded onto perhaps this planet" and that "it might be possible that one might find evidence for intelligent design if one looks at the details of biochemistry, molecular biology, where one might find a signature of some sort of designer". Now isn't that interesting? He used the word "signature", the same word used by Meyer as the title of his book. *The Signature in the cell*. I hardly think that this is coincidence, do you?

Dawkins is a very clever man, which I acknowledge, but only clever in the way he ducks and weaves when evidence comes his way that contradicts his world view. As new science emerges he cleverly prepares the ground beforehand so that should his pronouncements be found wanting he can fall back to a safe position and argue that he did at least consider possibilities that may differ from his previous world view.

Just in case I am accused of taking Dawkins words out of context, here is a transcript of what he said concerning Intelligent Design with Ben Stein.

Ben Stein: Well then who did create the heavens and the earth?
Richard Dawkins: Why do you use the word 'who'? You see you immediately beg the question by using the word 'who'.
Ben Stein: Well then how did it get created?
Richard Dawkins: Well, um, by a very slow process.
Ben Stein: Well how did it start?
Richard Dawkins: Nobody knows how it started. We know the kind of event that it must have been. We know the sort of event that must have happened for the origin of life.
Ben Stein: And what was that?
Richard Dawkins: It was the origin of the first self-replicating molecule.
Ben Stein: Right and how did that happen?
Richard Dawkins: I've told you, we don't know.
Ben Stein: So you have no idea how it started.
Richard Dawkins: No, no, nor has anyone.
Ben Stein: Nor has anyone else.
Ben Stein: What do you think is the possibility that Intelligent Design might turn out to be the answer to some issues in genetics or in Darwinian evolution?
Richard Dawkins: Well it could come about in the following way. It could be that, eh, at some earlier time somewhere in the universe a civilization evolved by probably some kind of Darwinian means to a very, very, high level of technology and designed a form of life that they seeded onto perhaps this planet. Ehm, now, that is a possibility and an intriguing possibility and I suppose it's possible that you might find evidence for that if you look at the um detail, details, of biochemistry, molecular biology, you might find a signature of some sort of designer.
Ben Stein: (voiceover, not part of interview) Wait a second,

Richard Dawkins thought Intelligent Design might be a legitimate pursuit.

Richard Dawkins: Um..and that designer could well be a higher intelligence from elsewhere in the universe.

Ben Stein: But, but

Richard Dawkins: But that higher intelligence would itself have had to have come about by some explicable, or ultimately explicable process, he couldn't have just jumped into existence spontaneously, that's the point.

Ben Stein: voiceover) So Professor Dawkins was not against Intelligent Design, just certain types of Designers, such as God. [32]

Well, there you have it - 'if you look at the details of biochemistry, molecular biology, you might find a signature of some sort of designer... and that designer could well be a higher intelligence from elsewhere in the universe'. In a nutshell, that's Intelligent Design no matter which way you may look at it.

Sadly, Dawkins and his ilk are stuck in a mentality of self-denial and I think even if God himself was to tap him on the shoulder and say, "here I am", I am pretty sure Dawkins would simply put his head in the sand and pretend he was having a hallucination. So what is it that can possibly convince people like Dawkins that what he sees as the appearance of design is in fact evidence of intelligent design? Will this book about Darwin's enigma about human hairless convince him? I doubt it? Why? I am afraid that only he can answer that question, but let us try to get inside his head on this matter.

Even Dawkins cannot help but acknowledge that both Darwin and Wallace admitted that natural selection could not account for human hairlessness. Since they were originators of the theory Dawkins can hardly oppose them after all the things that he has written about as the superiority of natural selection in his books over creation has been his mission in life. So that is a positive move forward, but what about Darwin's other theory of sexual selection? Dawkins has tried to explain this but as we have seen in an earlier chapter his explanation borders on the ridiculous. Nobody with a reasonable degree of intelligence could possibly subscribe to any of Dawkins' explanations on this matter. Let's hear what he has to say again and this time look pass the flannel!

*"The essential point is that male appearance and female taste evolve together in a kind of explosive chain reaction. ...**The ancestors of the peahens happened to take a step in the direction of preferring a larger fan**. It kicked in and,*

*within a very short time by evolutionary standards, peacocks were sprouting larger and more iridescent fans, and **females couldn't get enough of them.**"* (*The Ancestor's Tale: A Pilgrimage to the Dawn of Life*)

Darwin wrote in his *Descent of Man*

*"Many will declare that it is utterly incredible that a female bird should be able to appreciate fine shading and exquisite patterns. **It is undoubtedly a marvellous fact that she should possess this almost human degree of taste.** He who thinks that he can safely gauge the discrimination and taste of the lower animals may deny that the female Argus pheasant can appreciate such refined beauty; but he will then be compelled to admit that the extraordinary attitudes assumed by the male during the act of courtship, by which the wonderful beauty of his plumage is fully displayed, are purposeless; and this is a conclusion which I for one will never admit."*

Darwin declares that female peacock's taste is "a marvellous fact" and Dawkins says that they preferred (dictionary: more desirable than another) male "larger fan" and that "females couldn't get enough of them." Female taste was not a fact in Darwin's day nor is it today and Dawkins statement that females could not get enough of them is ridiculous.

The beauty of the peacock's tail as a sexual hook for female peacocks which is offered as evidence to support the sexual selection theory has since been shown to be nothing more than conjecture and not founded on any reliable scientific test. No wonder Darwin felt sick when thinking about the peacock tail, or should I day his tale, because he knew that what he was talking about was a lot of bull. Three recent scientific tests have proved the fallacy of Darwin's argument and it was one of the reasons why Wallace rejected Darwin's idea of sexual selection. Wallace said that the drab female peacock did not exhibit the same emotions of beauty that we humans share and he was right, as science has since proven.

With female taste used as an argument for sexual selection found wanting what can one make of Dawkins extension of this in connection with human hairlessness? Here is what he says in his book *The Ancestors's Tale*:

"Darwin believed that ancestral men found hairy women unattractive. Generations of men chose the most naked women as mates. Nakedness in men was dragged along in the evolutionary wake of nakedness in women, but never quite caught up, which is why men remain hairier than women. For

Darwin, the preferences that drove sexual selection were taken for granted - given. Men just prefer smooth women, and that's that."

Men just prefer smooth women, and that's that! What an extraordinary statement to make. Today, it is not likely that one will not find a smooth woman but Dawkins is really talking about our male hairy ape-like ancestors coming onto less hairy females. Dawkins is talking rubbish of course because there is certainly no evidence of this in the world of the great apes - chimpanzees, bonobos, gorillas and orangutans. In the case of the chimpanzee, our so called closest relative, males simply line up to wait their turn when a female comes on heat. She does not differentiate as to which males she mates with or how many males impregnate her as long as a result she becomes pregnant. During her state or period of heightened sexual arousal and activity she will mate pretty much non-stop until she comes out of it.

Perhaps Dawkins knew that he was treading on thin ice when he wrote his words about men preferring smooth women, because in the same book he follows up with the Parasite theory (Pagel and Bodmer) and Thermoregulation theory (Dr. Peter Wheeler), but only in passing as interesting possibilities, and he does not endorse either of them. This is understandable, because those theories do not stand up to proper scientific scrutiny either. So in this Dawkins has been wise to walk away, preferring instead be safe and go with his mentor Darwin's theory of sexual selection, while still not being able to put forward an argument that satisfies the criteria of good science. Can Dawkins really subscribe to Darwin's theory with the words, "For Darwin, peahens choose peacocks simply because, in their eyes, they are pretty?"

I CHALLENGE RICHARD DAWKINS

As it stands, there is no evolutionary mechanism that can account for man's hairlessness but there is a theory that does. It is quite simple really. We were created this way. Of course Dawkins and his ilk will not even consider this, such is their anti-God mindset.

Where does this leave us? It is now that I throw down the gauntlet and challenge Richard Dawkins or any other evolutionist for that matter, to explain a miracle of evolution. You see, our story does not finish here. There is one other problem that has been avoided by Neo-Darwinists like the plague which makes the issue regarding Man's almost hairless condition merely a sideshow.

"If it could be demonstrated that any complex organ existed, which could not possibly have been formed by numerous, successive, slight modifications, my theory would absolutely break down. But I can find out no such case."

Charles Darwin

"But my belief in evolution is not fundamentalism, and it is not faith, because I know what it would take to change my mind, and I would gladly do so if the necessary evidence were forthcoming."

Richard Dawkins

I would like to have challenged Charles Darwin because of what he declared in his famous *Origins of Species* book:

> "*If it could be demonstrated that any complex organ existed, which could not possibly have been formed by numerous, successive, slight modifications, my theory would absolutely break down. But I can find out no such case.*" [33]

Darwin did not need to look far for such an organ, but he is long since dead so I therefore I am unable to challenge him. But I can direct my challenge to his modern and most loyal advocate, Richard Dawkins. He too has made a similar statement as that of his mentor.

> "*But my belief in evolution is not fundamentalism, and it is not faith, because I know what it would take to change my mind, and **I would gladly do so if the necessary evidence were forthcoming.***" [34] [bold mine]

And to show that Dawkins is singing from the same song sheet as Darwin, in his book *The Blind Watchmaker*, he reiterates Darwin's premise that any incredulity about any incredibly complex transformation can be explained "**only if we stress that there was an extremely large number of steps along the way, and if each step is very tiny**". [35] [bold mine]

Well Mr Darwin and Mr Dawkins! There is a complex organ that could not be possibly have been formed by numerous, successive, slight modifications. In fact it would be a miracle if this was possible, but Dawkins does not believe in miracles, does he? But, with respects to this organ, he is prepared to believe in this one?

It is the purpose of a sequel to the present work entitled *Richard Dawkins' Miracle* that I bring Dawkins to task over his claim that man evolved from a common ancestor about six million years ago, the same ancestor that also spawned the chimpanzees and bonobo apes. This supposition is central to the entire theory of evolution with respects to the evolution of man and without it the theory would collapse like a pack of cards. It is the organ that I have alluded to that is discussed in great detail in the sequel which proves beyond a shadow of doubt that man could not have evolved from a hairy ape-like ancestor. Hence, in Darwin's own words, "*my theory would absolutely break down*", and it does big time.

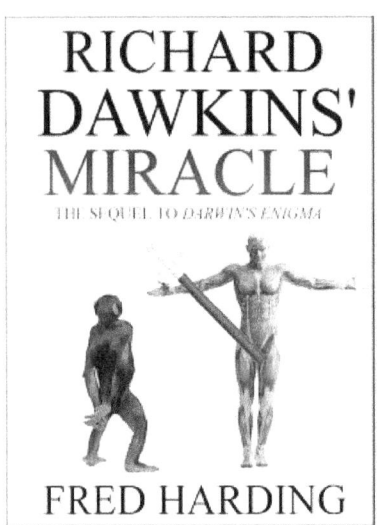

THE CHALLENGE!

"If it could be demonstrated that any complex organ existed, which could not possibly have been formed by numerous, successive, slight modifications, my theory would absolutely break down. But I can find out no such case."

Charles Darwin (1859)

"But my belief in evolution is not fundamentalism, and it is not faith, because I know what it would take to change my mind, and I would gladly do so if the necessary evidence were forthcoming."

Richard Dawkins (2007)

The question is this. Upon reading the present work and its sequel, will Dawkins see the error of his ways and acknowledge that he has been wrong, as his fellow atheist Antony Flew admitted to have been. Somehow, I doubt it, but who knows - miracles have been known to happen. Watch this space...

*** THE END ***

NOTES AND REFERENCES

PREFACE

[1] Thomas Henry Huxley, "Evidence as to Man's Place in Nature", 1863, p81
[2] Ibid; (p85)
[3] Ibid; (p86)
[4] Ibid; (p86)
[5] Ibid; (p125)
[6] "On the Law Which Has Regulated the Introduction of New Species", (S20: 1855)
[7] Ibid;
[8] Sanchez Manning, "Natural selection - it takes two: Darwin's rival Alfred Russel Wallace recognised at last", The Independent, 20 January 2013
[9] Ibid;
[10] Ibid;
[11] Ibid;
[12] Alfred Russel Wallace, "Contributions to the theory of natural selection", New York: Macmillan and co., 1871

Chapter 1 - NATURAL SELECTION IN CONFLICT

[1] Thomas Gilby, "St. Thomas Aquinas Philosophical Texts", Oxford University Press, 1951
[2] Carolus Linnaeus, The Fundamenta Botanica, No. 157 (1736)
[3] Several contemporary sources indicate that belief in species fixity was dominant at this time. In addition to Darwin's own comments in On the Origin of Species, Asa Gray described fixity as the "ordinary and generally received view" in the March 1860 issue of the American Journal of Science and Arts. The ninth edition of Charles Lyell's Principles of Geology (1853) argues for fixity, as does William Whewell's History of the Inductive Sciences (1858). The famed American scientist Louis Agassiz, as well as Scottish engineer Fleeming Jenkin, argued against Darwin from a fixist position.

[4] Charles Darwin, "On the Origin of Species by Means of Natural Selection, or the Preservation of Favoured Races in the Struggle for Life", London: John Murray, 1859, pp 3-4

[5] Russell Grigg, "Darwinism: it was all in the family". Much of his article is based on King-Hele, D., "Erasmus Darwin", Charles Scribner's Sons, New York, 1963

[6] "Jean-Baptiste Lamarck (1744-1829)", Museum of California Museum of Paleontology, Berkeley http://www.ucmp.berkeley.edu/history/lamarck.html

[8] Francis Darwin, "More Letters of Charles Darwin", D. Appledon, 1903

[9] Ibid;

[10] Robert Chambers, "Vestiges of the Natural History of Creation", London: John Churchill, 1844. p231

[11] Alexander Ireland, "Introduction to the Twelfth Edition," in Vestiges of the Natural History of Creation (1884), pp.vii-viii. Two years after the initial publication, in 1846, a Dr. Neil Arnott was also added to this tight inner circle.

[12] John van Wyhe, "The History of Phrenology on the Web", 2011. Wyhe is a historian of science, Senior Lecturer in the Departments of Biological Sciences and History and a Fellow of Tembusu College at the National University of Singapore. He is the Director of Darwin Online and Wallace Online, Professorial Fellow of Charles Darwin University, Fellow of the Linnean Society of London and a Scientific Associate of the Natural History Museum (London).

[13] Janet Brown, "Charles Darwin: Vol. 1 Voyaging", London: Jonathan Cape, 1995

[14] Letter 804 - Hooker, J. D. to Darwin, C. R., 30 Dec 1844". Darwin Correspondence Project.

[15] Letter 1082 - Darwin, C. R. to Hooker, J. D., (18 Apr 1847), Darwin Correspondence Project

[16] Charles Darwin, "On the Origin of Species by Means of Natural Selection, or the Preservation of Favoured Races in the Struggle for Life", London: John Murray, 1859, pp 3-4

[17] Robert Chambers, "Vestiges of the Natural History of Creation", 11th ed., London: John Churchill, 1860

[18] Darwin, Charles (1861), On the Origin of Species by Means of Natural Selection, or the Preservation of Favoured Races in the Struggle for Life, 3rd ed., London: John Murray

[19] Ibid; [20] Francis Darwin, "More Letters of Charles Darwin", D. Appledon, 1903

[21] Ibid;

[22] Burkhardt F. and Smith S. "The correspondence of Charles Darwin", 1982. Cambridge, vol. 11, p. 223.

[23] Janet Brown, "Charles Darwin: Vol. 1 Voyaging", London: Jonathan Cape, 1995

[24] Edward Larson, "Evolution: The Remarkable History of Scientific Theory", 2004.

[25] Michael Flannery, Associate Director for Historical Collections at the Lister Hill Library of the Health Sciences, and a professor at University of Alabama at Birmingham (UAB) http://www.alfredwallace.org/

[26] Michael Flannery, "Alfred Wallace - A Rediscoverd Life", 2011

Chapter 2 - DARWIN'S SECOND THEORY

[1] Charles Darwin, "The Descent of Man, and Selection in Relation to Sex", London: John Murray, 1871, p58

[2] Alfred Russel Wallace, "Contributions to the theory of natural selection", New York: Macmillan and co., 1871

[3] Charles Darwin, "On the Origin of Species by Means of Natural Selection, or the Preservation of Favoured Races in the Struggle for Life", London: John Murray, 1859

[4] Charles Darwin, "The Descent of Man, and Selection in Relation to Sex", London: John Murray, 1871, p58

[5] Ibid;

[6] Ibid;

[7] Charles Darwin, "On the Origin of Species by Means of Natural Selection, or the Preservation of Favoured Races in the Struggle for Life", London: John Murray, 1859

[8] Charles Darwin, "Letter 5475 to William Bernard Tegetmeier, editor, journalist, lecturer and naturalist. Pigeon-fancier and expert on poultry. Pigeon and poultry editor of the Field, 1864-1907. Secretary of the Apiarian Society of London, 30 March, 1867.

[9] Charles Darwin, "The Descent of Man, and Selection in Relation to Sex", London: John Murray, 1871

[10] Ibid;

[11] Peace Bella, "Lions Mating", Animal Kingdom. See also "Social Behaviour", College of Biological Sciences, University of Minnesota, and

[12] "Natural Selection", Dictionary.com

[13] Charles Darwin, "On the Origin of Species by Means of Natural Selection, or the Preservation of Favoured Races in the

Struggle for Life", London: John Murray, 1859

[14] Charles Darwin, "The Descent of Man", John Murray, London, p. 412, 1888

[15] Charles Darwin, "The Descent of Man, and Selection in Relation to Sex", London: John Murray, 1871

[16] Francis Darwin, "Letter to Asa Gray", dated 3 April 1860, "The Life and Letters of Charles Darwin",, D. Appleton and Company, New York and London, Vol. 2, pp. 90-91, 1911

[17] Charles Darwin, "The Descent of Man, and Selection in Relation to Sex", London: John Murray, 1871 [18] Takahashi, M., Arita, H., Hiraiwa-Hasegawa, M. & Hasegawa, T, Animal Behaviour 75, 1209-1219 (2008).

[19] Dakin, R. & Mongomerie, R. Animal Behaviour doi:10.1016/j.anbehav.2011.03.016 (2011).

[20] Jessica Yorzinski, "Animal attraction: peacocks and sexual selection", Journal of Experimental Biology, Positive Science, August 2013. Also, Washington Post, 2013-08-12.

[21] Charles Darwin, "The Descent of Man, and Selection in Relation to Sex", London: John Murray, 1871

[22] Ibid;

[23] Ibid;

[24] Ibid;

[25] Ibid;

[26] Ibid;

[27] Dr Alun Withey, "Beards, Moustaches and Facial Hair in History", 26 April 2012

[28] "Hairy Men: Are They An Instant Turn On?", likecosmetics.com

[29] Alfred Wallace, "Darwin's "The Descent of Man and Selection in Relation to Sex", The Academy, 15 March 1871

[30] Alfred Wallace, "Darwinism", 1889

[31] Thomas Hunt Morgan, "Evolution and Adaptation", 1903, New York: Macmillan

[32] Max Westenhöfer, " Der Eigenweg des Menschen. Dargestellt auf Grund von vergleichend morphologischen Untersuchungen über die Artbildung und Menschwerdung. Verlag der Medizinischen Welt, W. Mannstaedt & Co., Berlin. (1942)

Chapter 3 - THE AQUATIC APE

[1] Max Westenhöfer, "Der Eigenweg des Menschen (The Unique Road to Man)", W. Mannstaedt & Co., Berlin, 1942.

[2] Ibid;

[3] Elaine Morgan, "The Naked Darwinist", Eildon Press, 2008

[4] Ibid;

[5] Alister Clavering Hardy, "Was man more aquatic in the past", New Scientist. 26 March, 1960 7: 642-645.

[6] The New Scientist, 17 March 1960

[7] Ibid;

[8] Ibid;

[9] "J. S. Weiner and the exposure of the Piltdown forgery", Antiquity March 1983

[10] Elaine Morgan, "The Naked Darwinist", Eildon Press, 2008

[11] Ibid;

[12] Ibid;

[13] Elaine Morgan, "Aquatic Ape Theory" primitivism.com

[14] Ibid;

[15] Ibid;

[16] Ibid;

[17]Ibid;

[18] Caroline Pond, "Not an aquatic ape -- just an exceptionally fat mammal", New Scientist 4:64-65. p65. Caroline Pond Ph.D., is a Senior Lecturer in Biology at the Open University in Milton Keynes

[19] Ibid;

[20] "The Aquatic Ape: Fact or Fiction", Proceedings from the Valkenburg Conference, Souvenir Press. (1991) 369 pages. http://www.riverapes.com/AAH/FoF/FactOrFiction.htm

[21] Ibid;

[22] Timur Moon, "Experts Pour Cold Water on David Attenborough 'Aquatic Ape' Claims", The Guardian, 27th April, 2013

[23] Ibid;

[24] John H. Langdon, "Umbrella hypotheses and parsimony in human evolution: a critique of the Aquatic Ape Hypothesis", Journal of Human Evolution Volume 33, Issue 4, October 1997, Pages 479-494

Chapter 4 - NOW IT GET'S HOT

[1] GD Ruxton, DM Wilkinson, "Thermoregulation and endurance running in extinct hominins: Wheeler's models revisited", Journal of Human Evolution, August 2011

[2] Mariette DiChristina, "Continuum of Change: The Hairless

Human", Scientific American, February 2010 issue

[3] Ken Harding, evolution.mbdojo.com/you_figure_it_out.htm

[4] Ann Gibbons, "The Human Family's Earliest Ancestors", The Smithstonian Magazine, March 2010

[5] Gary Parker, "Building Blocks in Science", New Leaf Publishing, 2007, p127

[6] Christine Berge, "How did the australopithecines walk? A biomechanical study of the hip and thigh of Australopithecus afarensis", Journal of Human Evolution, Volume 26, Issue 4, April 1994, Pages 259-273

[7] Charles Choi, "Early Human 'Lucy' Swung from the Trees", LiveScience October 25, 2012

[8] Ruxton GD, Wilkinson DM, "Thermoregulation and endurance running in extinct hominins: Wheeler's models revisited", Journal of Human Evolution, 2011 Aug;61(2):169-75.

[9] Homo erectus, "Natural History Museum", 2013

[10] Ibid;

[11] Charles Darwin, "The Descent of Man, and Selection in Relation to Sex", London: John Murray, 1871

Chapter 5 - DESPERATE FANTASIES

[1] Charles Darwin, "The Descent of Man, and Selection in Relation to Sex", London: John Murray, 1871

[2] John Pickrell, "Are Humans Furless to Thwart Parasites?", National Geographic News, June 17, 2003 See also: Pagel M., Bodmer, W., "A naked ape would have fewer parasites" Proceedings of the Royal Society B: Biological Sciences, 270 (2003) S117 - S119, DOI 10.1098/rsbl.2003.0041

[3] Animal Planet, "Chimpanzees", Discovery Channel.

[4] Bentley Glass, "Evolution of Hairlessness in Man", Science 15 April 1966: 294.

[5] Alan Rogers, David Iltis, Stephen Wooding, Genetic Variation at the MC1R Locus and the Time since Loss of Human Body Hair", Current Anthropology, 2004

[6] Karen Radner, Eleanor Robson, "The Oxford Handbook of Cuneiform Culture", p672

[7] Morgan, E. (1982). The aquatic ape. London: Souvenir Press.

[8] Markus J. Rantala, "Evolution of nakedness in Homo sapiens", Journal of Zoology, 28 November 2006

[9] Shannen L Robson, Bernard Wood, "Hominin life history: reconstruction and evolution", Jornal of Anatomy. 2008 April; 212(4): 394-425.

[10] Markus J. Rantala, "Evolution of nakedness in Homo sapiens", Journal of Zoology, 28 November 2006. See also "The Descent of Madness: Evolutionary Origins of Psychosis and the Social Brain" by Jonathan Burns

[11] Kenneth McNamara, "Shapes of Time: The Evolution of Growth and Development", The Johns Hopkins University Press, 1997

[12] Gary G. Schwartz, Leonard A. Rosenblum, "Allometry of primate hair density and the evolution of human hairlessness", American Journal of Physical Anthropology Volume 55, Issue 1, pages 9-12, May 1981

[13] William Stephensson, "The ecological development of man.", Sydney: Angus and Robertson, 1972

[14] Bernard Campbell, "Human evolution", Chicago: Aldine. 1966

Chapter 6 - THE DNA GAME

[1] Richard Dawkins, "The Blind Watchmaker: Why the Evidence of Evolution Reveals a Universe Without Design", W.W. Norton, New York, p. 263, 1986.

[2] John Pickrell, "Humans, Chimps Not as Closely Related as Thought?", National Geographic News, September 24, 2002

[3] Izabela Depczyk, "Polish scientists closer to pig-to-human body-part transplants", Warsaw Business Journal, quoting professor Zdzislaw Smorag

[4] John Hewitt, "A chimp-pig hybrid origin for humans?", Physorg, July 03, 2013

[5] Ibid;

[6] John Hewitt, "Human hybrids: a closer look at the theory and evidence", Physorg, July 25, 2013

[7] "An Overview of the Human Genome Project", National Human Genome Research Institute, http://www.genome.gov/12011238

[8] "The Power of Sequencing Single Cell Genomes", National Human Genome Research Institute, http://www.genome.gov/27552295

[9] The Chimpanzee Sequencing and Analysis Consortium, "Initial sequence of the chimpanzee genome and comparison with the human genome", Nature, 1 September, 2005

[10] Ibid;

[11] Ibid;

[12] Rick Weiss, "Scientists Complete Genetic Map of the

Chimpanzee", The Washington Post, 1 September, 2005

[13] "New Genome Comparison Finds Chimps, Humans Very Similar at the DNA Level", National Human Genome Research Institute, http://www.genome.gov/15515096

[14] Ibid;

[15] Richard Dawkins, "The Greatest Show on Earth", 2009, pp. 332-333

[16] Alok Jha, "Breakthrough study overturns theory of 'junk DNA' in genome", The Guardian, 5 September 2012

[17] Ibid; [18] King M-C and Wilson AC "Evolution at two levels in humans and chimpanzees", Science 188 107-116

[19] Michel Morange, "The genetic distance between humans and chimpanzees What did Mary-Claire King and Allan Wilson really say in 1975?", Journal of Biosciences, 2011

[20] Mary-Claire King, "Shared genetic material between humans and chimps", DNA Learning Center, Cold Spring Harbor Laboratory. ID 15119

[21] King M-C and Wilson AC "Evolution at two levels in humans and chimpanzees", Science 188 107-116

[22] J R. Minkel, "Human-Chimp Gene Gap Widens from Tally of Duplicate Genes", Scientific American, December 19, 2006. Based upon a report in the "Proceedings of the National Academy of Sciences" (DOI: 10.1073/pnas.172510699)

[23] Lawrence B. Schook, "Swine Genome Sequencing Consortium (SGSC): a strategic roadmap for sequencing the pig genome", Comparative and Functional Genomics, 2005; 6: 251-255.

[24] Buchen, L., "The fickle Y chromosome", Nature 463 (7278):149, 14 January 2010.

[25] Ibid;

[26] Ann Gibbons, "Y Chromosome Evolving Rapidly', Science, 13 January 2010

[27] John Hawks, "Unbelievable Y chromosome differences between humans and chimpanzees", 2010-01-14 00:11 Referring to Jenniver Hughes and 16 others. "Chimpanzee and human Y chromosomes are remarkably divergent in structure and gene content.", Nature (early online) doi:10.1038/nature08700, Nature 463, 536-539 (28 January 2010) | doi:10.1038/nature08700; Received 3 August 2009; Accepted 24 November 2009; Published online 13 January 2010

[28] Jeffrey P. Tomkins, "Comprehensive Analysis of Chimpanzee and Human Chromosomes Reveals Average DNA Similarity of 70%", Answers Research Journal 6 (2013): 63-69.

[29] Jon Cohen,"Relative differences: the myth of 1%", Science 316 1836, 2007

[30 Joan U. Pontius, James C. Mullikin, Douglas R. Smith, Agencourt Sequencing Team, Kerstin Lindblad-Toh, Sante Gnerre, Michele Clamp, Jean Chang, Robert Stephens, Beena Neelam, Natalia Volfovsky, Alejandro A. Schäffer, Richa Agarwala, Kristina Narfström, William J. Murphy, Urs Giger, Alfred L. Roca, Agostinho Antunes, Marilyn Menotti-Raymond, Naoya Yuhki, Jill Pecon-Slattery, Warren E. Johnson, Guillaume Bourque, Glenn Tesler, NISC Comparative Sequencing Program, and Stephen J. O'Brien, "Initial sequence and comparative analysis of the cat genome", Genome Research, 2007. 17: 1675-1689

[31] John Hawkins, "Unbelievable Y chromosome differences between humans and chimpanzees", John Hawkins Weblog, 14 January 2010

[32] Ibid;

[33] Hughes JF and 16 others. "Chimpanzee and human Y chromosomes are remarkably divergent in structure and gene content." Nature doi:10.1038/nature08700, 2010

[34] John Hawkins, "Unbelievable Y chromosome differences between humans and chimpanzees", John Hawkins Weblog, 14 January 2010

[35] Ibid;

[36] "Initial sequence and comparative analysis of the cat genome", Genome Research November 2007, 17(11): 1675-1689.

Chapter 7 - FROM THE ASHES OF THE PHOENIX

[1] Paulo Gama Mota, "Darwin's sexual selection theory - a forgotten idea", Departamento de Ciências da Vida, Universidade de Coimbra, Portugal

[2] Alfred Wallace, "To-day's centenaries: Charles Darwin", Daily Mail No. 4007 (12 February 1909)

[3] Z. Tang-Marti´nez, "Bateman's Principles: Original Experiment and Modern Data For and Against", 2010

[4] Leslie Real, Behavioral Mechanisms in Evolutionary Ecology, The University of Chicago Press 1994, p. 5

[5] Olivia Judson, "Dr. Tatiana's Sex Advice to All Creation: The Definitive Guide to the Evolutionary Biology of Sex", Henry Holt 2002, p. 13

[6] "Biologists Reveal Potential 'Fatal Flaw' in Iconic Sexual Selection Study", Science News, 26 June 2012. This study was

federally funded by the National Science Foundation. Patricia Adair Gowaty quoted. Other co-authors included Wyatt Anderson, a professor of genetics at the University of Georgia and a member of the National Academy of Sciences, and Yong-Kyu Kim, a research scientist at Emory University. The original article was written by Kim DeRose.

[7] Ibid;
[8] Andreas Paul, "Sexual Selection and Mate Choice", International Journal of Primatology, Vol. 23, No. 4, August 2002 (°C 2002)
[9] Paulo Gama Mota, "Darwin's sexual selection theory - a forgotten idea", Universidade de Coimbra, Portugal
[10] Michel Ohmer, "Challenging Classic Sexual Selection Theory: The baby became the bathwater years ago, but no one noticed until now", John S Knight Institute for Writing in Disciplines, Cornell University, New York, Spring 2008, No 9
[10] Irwin Bernstein, "The study of things I have seen", American Journal of Primatology: 60:77-84 2003
[11] Ibid;
[12] Tim Clutton-Brock, "Sexual Selection in Males and Females", Science 318, 2007
[13] Richard Dawking, "The God Delusion", I once wrote that anybody who didn't believe in evolution must be stupid, insane or ignorant, and I was then careful to add that ignorance is no crime. I should now update my statement. Anybody who doesn't believe in evolution is stupid, insane, or hasn't read Jerry Coyne. I defy any reasonable person to read this marvellous book and still take seriously the "breathtaking inanity" that is intelligent design "theory" or its country cousin, young earth creationism.
[14] Martin Beckford, "Richard Dawkins branded 'secularist bigot' by veteran philosopher", The Telegraph, 2nd August 2008

[15] Mark Stuertz, "God in the Details" Dallas Observer News, May 3 2007
[16] Ian Sample, "Martin Rees wins controversial £1m Templeton prize", The Guardian, 6 April 2011
[17] Mark Vernon, "Martin Rees's Templeton prize may mark a turning point in the 'God wars'", The Guardian, 6 April 2011
[18] Ibid;
[19] Tim Ross, "Richard Dawkins accused of cowardice for refusing to debate existence of God", Daily Telegraph, 14 May 2011
[20] Stephen Pollard, "For once, Richard Dawkins is lost for

words", Daily Telegraph, 14 February 2012.

[21] Sam Jones, "Richard Dawkins is the phall guy at Cambridge debate with Williams" - Britain's most ardent atheist lost the encounter with former archbishop on paper - but he got laughs for his penis joke, The Guardian, Friday 1 February 2013

[22] Alok Jha, "Peter Higgs criticises Richard Dawkins over anti-religious 'fundamentalism: Higgs boson theorist says he agrees with those who find Dawkins' approach to dealing with believers 'embarrassing", The Guardian, 26 December 2012

[23] Mark Wallace, "Richard-dawkins-an-embarrassment-to-atheists", 24 August 2010, (crashbangwallace.com)

[24] Martin Robbins, "Atheism is maturing, and it will leave Richard Dawkins behind", NewStatesman, 9 August, 2013

[25] Terry Eagleton, "Lunging, Flailing, Mispunching", London Review of Books, Vol. 28 No. 20 · 19 October 2006 pages 32-34

[26] Ronald Fisher, "The Genetical Theory of Natural Selection", Oxford: Clarendon Press, 1930.

[27] J. R Krebs, N. B Davies, "Behavioural Ecology: An Evolutionary Approach", Blackwell Pubilishing, 1997. N.Davies is Professor of Behavioural Ecology at the University of Cambridge.

[28] Malte Andersson, "Sexual Selection", Princeton University Press, 1994

[29] [1] Charles Darwin, "The Descent of Man, and Selection in Relation to Sex", London: John Murray, 1871

[30] Geoffrey F. Miller, "Evolution of human music through sexual selection", 2001

[31] Joseph Jordania, "Why do People Sing? Music in Human Evolution.", Logos, 2011:186-196.

[32] Razib Khan, "Sexual selection: ignore the blonde?", Discover, 23 June, 2013

[33] Ibid;

[34] Charles Darwin, "Letter 431 to A.R. Wallace", March 19th, 1868.

[35] Alfred Wallace, "Review of The Descent of man by Darwin", Academy 2 (no. 20, 15 March, 1871): 177-183.

[36] Thomas Hunt Morgan, "Evolution and Adaptation". Macmillan & Co., Ltd, 1908

Chapter 8 - DARWIN'S ENIGMA CONTINUES

[1] Richard Dawkins, "Big ideas: Evolution", New Scientist, 17 September 2005

[2] Alfred Russel Wallace, "Contributions to the theory of natural

selection", New York: Macmillan and co., 1871

[3] Charles Darwin, "The Descent of Man and Selection in Relation to Sex", John Murray, 1874

[4] Ibid;

[5] Gowaty, Kim, and Anderson, "Mendel's law reveals fatal flaws in Bateman's 1948 study of mating and fitness", PMC Jan 1, 2013; 7(1): 28-38. - "No evidence of sexual selection in a repetition of Bateman's classic study of Drosophila melanogaster" in Proc Natl Acad Sci U S A, volume 109 on page 11740.

[6] Ibid;

[7] Michel Ohmer, "Challenging Classic Sexual Selection Theory: The baby became the bathwater years ago, but no one noticed until now", 2006

[8] Nick Atkinson, "Sexual selection alternative slammed Biologists write to Science to defend the theory of sexual selection", The Scientist, May 5, 2006

[9] Michel Ohmer, "Challenging Classic Sexual Selection Theory: The baby became the bathwater years ago, but no one noticed until now", 2006

[10] Mariko Takahashi, "Peahens do not prefer peacocks with more elaborate trains", Animal Behaviour, Volume 75, Issue 4, April 2008, Pages 1209-1219

[11] Elaine Morgan, "The Naked Darwinist", - "We are Furless (6), 2008

[12] Ibid;

[13] Francis Ebling, "Journal of Human Evolution", January 1985

[14] Richard Dawkins, "The Blind Watchmaker: Why the Evidence of Evolution Reveals a Universe Without Design", W.W Norton and Co: New York NY, 1986.

[15] Alison Abbott. "Pig geneticists go the whole hog: Genome will benefit farmers and medical researchers", Nature 14 November 2012.

[16] Tomkins, J. 2013. "Comprehensive Analysis of Chimpanzee and Human Chromosomes Reveals Average DNA Similarity of 70%", Answers Research Journal. 6 (2013): 63-69.

[17] Izabela Depczyk, "Polish Scientists Closer to Pig-to-Human body-part transplants", Warsaw Business Journal, Poland 13.02.2012

[18] Ross Pomeroy, "Baboons & Humans: Being Beta is Not So Bad", Real Clear Science, July 17, 2011

[19] Eugene McCarthy, "The Hybrid Hypothesis - A new theory

of human origins", Macroevolution.net

[20] Sullivan TP1, Eaglstein WH, Davis SC, Mertz P, "The pig as a model for human wound healing.", Wound Repair Regen. 2001 Mar-Apr;9(2):66-76. PMID: 11350644 [PubMed - indexed for MEDLINE]

[21] Mark Prigg, "How Animal Farm was right: Pigs really ARE almost identical to humans, say scientists", MailOnline, 14 November 2012

[22] "Doctors graft pig's skin onto burned child in Zigong", GoChengdoo News blog, February 12, 2012

[23] Eugene McCarthy, "The Hybrid Hypothesis - A new theory of human origins", Macroevolution.net

[24] Ibid;

[25] Ibid;

[26] John Hewitt, "A chimp-pig hybrid origin for humans?", July 03, 2013, Phys.org

[27] Thomas H. Huxley, " Evidence as to Man's place in nature", Williams & Norgate, London. p114-115, 1863

[28] John Hewitt, "Human hybrids: a closer look at the theory and evidence", July 26, 2013, Phys.org

[29] Ignazio de Vega, "In a Thing So Small", Review of "Signature in the Cell" by Stephen Meyer, HarperOne, 2009

[30] Ibid;

[31] Alastair Noble, "Richard Dawkins endorses Intelligent Design", Centre for Intelligent Design, December 2012.

[32] Ibid;

[33] Charles Darwin, "On the Origin of Species", 1859

[34] Alister McGrath, "The Dawkins Delusion", AlterNet, 25 January, 2007

[35] Richard Dawkins, "The Blind Watchmaker: Why the Evidence of Evolution Reveals a Universe Without Design", W.W Norton and Co: New York NY, 1986. , xvi + 332 pp.

Books by the Author